Knowledge Production in Mao-Era China

Knowledge Production in Mao-Era China

Learning from the Masses

Marc Andre Matten and Rui Kunze

LEXINGTON BOOKS
Lanham • Boulder • New York • London

Published by Lexington Books
An imprint of The Rowman & Littlefield Publishing Group, Inc.
4501 Forbes Boulevard, Suite 200, Lanham, Maryland 20706
www.rowman.com

86-90 Paul Street, London EC2A 4NE

Copyright © 2021 by The Rowman & Littlefield Publishing Group, Inc.

All rights reserved. No part of this book may be reproduced in any form or by any electronic or mechanical means, including information storage and retrieval systems, without written permission from the publisher, except by a reviewer who may quote passages in a review.

British Library Cataloguing in Publication Information Available

Library of Congress Cataloging-in-Publication Data

ISBN: 978-1-4985-8461-6 (cloth)
ISBN: 978-1-4985-8463-0 (paper)
ISBN: 978-1-4985-8462-3 (electronic)

Contents

List of Figures		vii
Acknowledgments		xi
Introduction		xiii
1	Defining Correct Science: Transformations of Knowledge Epistemologies	1
2	Creating the People's Science: Science Dissemination as a Social Process	29
3	Promising a Bright Future: The (Half-)Mechanization of Agricultural Production	59
4	Producing Knowledge on the Shopfloor: Technological Innovation in Socialist Industrialization	81
5	Creating a Bifurcated Knowledge System: The Case of Chinese Veterinary Medicine	105
6	Re-shuffling Science in the Reform Era	125
Bibliography		141
Index		161
About the Authors		167

Figures

COVER IMAGES

Cover Figure 01
Map of Acupuncture Points of a Pig (*Zhu zhenjiu xuewei tu* 猪针灸穴位图, designer unknown), in Jiangsu sheng nongye kexue yanjiusuo geming weiyuanhui 江苏省农业科学研究所革命委员会. *Zhongshouyi zhenliao xuanbian* 中兽医诊疗选编 (Nanjing: Jiangsu sheng geming weiyuanhui, 1970), annex.

Cover Figure 02
Mobilize the whole population, to make sure that steel production is doubled! (*Quanmin dongyuan, baozheng gangtie fanyifan!* 全民动员，保证钢铁翻一番! Designer unknown), Shanghai renmin meishu chubanshe (上海人民美术出版社, 1958), International Institute of Social History (Amsterdam). For more information about the copyright see the IISH copyright statement (iisg.amsterdam/en/collections/using/reproductions).

FIGURES

Figure 1.1
Zhongyang weishengbu weisheng xuanchuanchu 中央卫生部卫生宣传处, "Yufang xiaji chuanranbing 预防夏季传染病 (Preventing Contagious Diseases in the Summer Season)—Part I," *Kexue puji gongzuo*, no. 16 (June 1951): 179.

Figure 1.2
Zhongyang weishengbu weisheng xuanchuanchu 中央卫生部卫生宣传处, "Yufang xiaji chuanranbing 预防夏季传染病 (Preventing Contagious Diseases in the Summer Season)—Part II," *Kexue puji gongzuo*, no. 16 (June 1951): 180.

Figure 1.3
Zhongyang weishengbu weisheng xuanchuanchu 中央卫生部卫生宣传处, "Yufang xiaji chuanranbing 预防夏季传染病 (Preventing Contagious Diseases in the Summer Season)—Part III," *Kexue puji gongzuo*, no. 16 (June 1951): 181.

Figure 1.4
Zhongyang weishengbu weisheng xuanchuanchu 中央卫生部卫生宣传处, "Yufang xiaji chuanranbing 预防夏季传染病 (Preventing Contagious Diseases in the Summer Season)—Part IV," *Kexue puji gongzuo*, no. 16 (June 1951): 182.

Figure 2.1
Zhongyang renmin kexueguan 中央人民科学馆. *Kangmei yuanchao yundong zhong de dongbei yu chaoxian tuji* 抗美援朝运动中的东北与朝鲜图集 (Atlas of the Northeast and Korea in the Campaign of Resisting America and Aiding Korea). Shanghai: Zhonghua shuju, 1951.

Figure 2.2
A magnified view showing the jewel wasp fighting against the pink bollworm, still frame of the movie *The Jewel Wasp and the Pink Bollworm* (*Jinxiaofeng yu honglingchong* 金小蜂与红铃虫, 1963).

Figure 2.3
Cover page, *Knowledge Is Power* (*Zhishi jiushi liliang* 知识就是力量), no. 9 (1957).

Figure 3.1
Quanguo nongju zhanlanhui 全国农具展览会, ed. *Quanguo nongju zhanlanhui—tuijian zhanpin—nongtian paiguan jixie* 全国农具展览会—推荐展品—农田排灌机械 (National Exhibition of Agricultural Tools—Recommended Exhibits—Farmland Irrigation Machines). Beijing: Kexue puji chubanshe, 1958.

Figure 3.2
Quanguo nongju zhanlanhui 全国农具展览会, ed. *Quanguo nongju zhanlanhui—tuijian zhanpin—nongtian paiguan jixie* 全国农具展览会—推荐展品—农田排灌机械 (National Exhibition of Agricultural Tools—Recommended Exhibits—Farmland Irrigation Machines). Beijing: Kexue puji chubanshe, 1958.

Figure 4.1
Cover page, *Knowledge Is Power* (*Zhishi jiushi liliang* 知识就是力量), no. 12 (1958).

Figure 5.1
Map of Acupuncture Points of a Pig (*Zhu zhenjiu xuewei tu* 猪针灸穴位图, designer unknown), in Jiangsu sheng nongye kexue yanjiusuo geming weiyuanhui 江苏省农业科学研究所革命委员会. *Zhongshouyi zhenliao xuanbian* 中兽医诊疗选编 (Nanjing: Jiangsu sheng geming weiyuanhui, 1970), annex.

The images in this book are in the public domain. According to copyright laws of the People's Republic of China all images works, and all works whose copyright holders is a juristic person, enter the public domain 50 years after they were first published. Every effort has been made to contact the copyright holders. However, if there is anybody who has not been contacted, please contact the publishers in the first instance.

Acknowledgments

This book owes its existence first and foremost to an extraordinary book donation that our institute at Friedrich-Alexander University Erlangen-Nuremberg (FAU) received from the Shanghai Academy of Social Sciences (SASS) in 2006. Thanks to the initiative of Xiong Yuezhi, roughly 100,000 books and bounded journals arrived in Erlangen, representing the book production from the late 1940s to the early 1980s. Particularly fascinating are publications in the fields of economy, industry, agriculture and trade, science and technology, to the history of China and the world, as well as Chinese and foreign literature (for a detailed overview on the collection at FAU see www.sass.fau.de). Without the opportunity to browse through the stacks, flipping through popular publications and expert handbooks of the Mao era, this book would not have been possible.

We would like to express our special gratitude to the Chiang Ching-kuo Foundation for generously supporting our research project *Science, Modernity and Political Behavior in Contemporary China (1949–1978)* (RG007-U-12). The foundation financed not only the research of the project, but also the kick-off workshop "Production and Distribution of Medical Knowledge in Rural China: Maoist and Post-Maoist Experiences" at FAU in 2014.

For many productive discussions that helped us develop ideas, we are grateful to friends and colleagues: Chang Che-chia, Arunabh Ghosh, Renée Krusche, Tong Lam, Sean Hsiang-lin Lei, Sigrid Schmalzer, Wang Mingde, and the anonymous reviewers for their valuable comments.

Our thanks also go to our research assistants Jana Cyrol, Julia Hauser, Jonas Humpert, and Wang Mingde, who delved into and discovered gems in the SASS collection in the early phase of the project; and Luisa Kostka, Carolin Tzschentke, and Maren Wicher for their help with formatting and finalizing the manuscript.

Introduction

In 1951 the illustrated magazine *Science Pictorial* (*Kexue huabao* 科学画报) serialized the *Inventions of Our Ancestors* (*Women zuxian de chuangzao faming* 我们祖先的创造发明), an account of technological inventions, scientific development, and medical and healing practices since the ancient era. It was then published as a fourteen-chapter popular fiction in September 1951 and was revised and republished several times from 1957 to 1962.

The 1951 version, written in a style that facilitated its adaptation for storytelling performances, intended to communicate the idea of science (*kexue* 科学) and its history to the masses of socialist China in the entertaining form of tales.[1] Employing the Marxist-Leninist historical periodization, it attributed the country's weakness in modern science to despotic feudalism and the invasion of imperialism. China's five thousand years of written history shows, however, the author Mao Zuoben 茅左本 argued, that "we do not only have science, but our science has immensely influenced the formation of Western civilization."[2] While the fields of science and technology suffered corruption and waste of talents in Republican China (caused by the suppression of native knowledge, persecution of scientists, lack of occupation opportunities, etc.), the "new society" under Mao Zedong's 毛泽东 (1893–1976) leadership "returned the five-thousand-year [legacy of] science and technology to the hands of the laboring people (*laodong renmin* 劳动人民)," who were now eager to make full use of their talent and creativity to produce more glorious knowledge of science and technology.[3] The book did concede, on the other hand, that some legacy, such as Chinese medicine which "has been passed down from the feudal society and was based totally on experience (*jingyan* 经验)," should be "scientized" (*kexuehua* 科学化). To do so, practitioners of Chinese medicine should learn scientific methods and identify success-

xiii

ful Chinese medical pharmaceutics,[4] so that knowledge from the feudal past could survive in the modern era.

Like many other popular science materials in the early People's Republic of China (PRC), this 1951 book integrates the notion of "science" into the state-building and socialist construction (*shehuizhuyi jianshe* 社会主义建设). First, it cultivates a patriotism with a meaningful history of science by proudly positioning a Chinese genealogy of "science" side-by-side with modern science. Secondly, it periodizes China's "five-thousand-year history" in the historical framework of Marxism-Leninism, which is the state ideology whose evolutionary thinking claims itself to be "scientific." Thirdly, it identifies the emancipated "laboring people"—scientists as well as farmers and workers—as the new active agent of science and technology in the "new society." Last but not least the *Inventions of Our Ancestors* also praised the ideal of a classless society where science and technology were accessible to all social classes.

The attitude toward the relation of "science" to "experience" changed in the 1957 version and later reprints. Mao Zuoben added new research results achieved between 1951 and 1957,[5] provided more illustrations, and rewrote the book in the form of short essay. The new version is more readable, but less useful for storytelling performances. It perhaps assumed that after the literacy campaigns in the first half of the 1950s more people were able to read. In addition, instead of enumerating the scientific and technological achievements in ancient China the 1957 version recategorized its contents according to the taxonomy of modern science, such as mechanics, agricultural science, meteorology, metallurgy, medicine and pharmaceutics, etc. It self-consciously positioned the national achievements as scientific, and most notably Chinese medicine was presented in a much more positive light. The 1957 book no longer denounced such knowledge as feudal and superstitious practice as it had been done in the Republican era, and it even stopped demanding the need of "scientizing" Chinese medical legacy. Instead, the 1957 version praised Chinese medicine as "[the result of] experiences accumulated by our ancestors through their struggle against diseases over thousands of years" and "its rich contents have contributed tremendously to the health of humanity (我国的医学，是我们祖先数千年来和疾病斗争的经验积累，内容非常丰富，对人类的健康有着重大的贡献)."[6] It cheered the foundation of the Academy of Chinese Medical Sciences (*Zhongyi yanjiuyuan* 中医研究院) in 1955, the first state-founded institution of Chinese medicine. Following Mao Zedong's demand on the First National Conference on Public Health (*Diyi jie quanguo weisheng gongzuo huiyi* 第一届全国卫生工作会议) in 1950 to combine both new and old medicine and build a united front of medical workers in both Chinese and Western medicine, the academy taught experts in Western medicine Chinese medical knowledge.[7]

As a popular science book whose (re)publications spanned the turbulent years from 1951 to 1962, the substantial or minor modifications in different versions of the *Inventions of Our Ancestors* register the changing and sometimes contradicting official discourses on knowledge production, standards, and definitions of "science" as well as their associated political and social concerns. The increasing endorsement of empirical experience of the laboring people in the discourse of science, as we will discuss below, can be traced to Mao Zedong's epistemology of "practice" (*shijian* 实践) in knowledge production and evaluation, which underlies the "scientific" method of "experiment" (*shiyan* 试验) and emphasizes the experimenter's first-hand experience. The changing discourse of science would then foreground the role of non-experts in collecting and preserving knowledge, and when combined with class politics and pragmatic needs, would also open up the possibility of plural knowledge systems. The plurality of knowledge, as we argue in the course of the book, has important implications for understanding science discourse in Mao China as well as post-Mao China, especially with regard to its complex and continuously shifting relation with politics.

No doubt the *Inventions of Our Ancestors* is part of the state propaganda, which, in Sigrid Schmalzer's definition, is "material produced by the state specifically to promote the state's priorities." Schmalzer argues that they should nevertheless be taken seriously by historians, because they not only articulated the state's "visions of science" but also "meant a great deal to, and had very real consequences for, people on the ground."[8] In the Mao era the state propaganda materials on science included newspapers, magazines, posters, official reports, manuals of new technology and techniques, etc. They, to a large extent, shaped the public discourse—and public opinions—of "science" in that era.[9] Due to the fact that the Chinese Communist Party's (CCP) leadership had imposed its political will onto the public discourse of science, most infamously during the Great Leap Forward (1958–61) and Cultural Revolution (1966–76) years, theories of Maoist voluntarism—or more generally, political ideology intervening with and therefore hindering the development of science—have been used to explain the knowledge production in the Mao era.[10] Yet by presuming an opposition between autonomous science and political ideology, these theories forestall a productive examination of their interactive dynamics in the decades between the 1950s and the 1970s. By taking propaganda as an important factor, we hope to present a more nuanced analysis of the public discourse of science in the Mao era, which tries to explain the persistent faith of the Chinese populace in (the power of) science—from late Qing via Republican China and Mao-era China all the way to the present—to realize historical progress, national modernity, and individual well-being, despite all the ambiguities, contradictions, disputes, and negotiations in describing and prescribing what "science" is and should be.

THE DISUNITY OF SCIENCE: SITUATING SCIENCE IN MAO-ERA CHINA

In recent years, historians of science and technology have argued that there has never been a unified, consistent notion of science.[11] Not only the aims of science, scientific methods, and evaluative strategies for scientific propositions, but also the taxonomies of scientific disciplines have varied in different historical and cultural contexts, and are still contested today.[12] Experiment, which has often been considered by the public as "just another name for scientific method," for example, may have different goals—"testing a theory about the phenomena under scrutiny" or "simply out of curiosity to see what will happen."[13] And in Chinese, "experiment" with these two goals appears as two words: *shiyan* 实验 for the former, while *shiyan* 试验 for the latter. Karin Knorr-Cetina shows that experimental practices can be drastically different for scientists in the high-energy physics laboratory and for those in a molecular biology laboratory.[14] In addition, when scientific knowledge is communicated from the expert to the lay public, Terry Shinn and Richard Whitley argue, redescription and translation take place, which alter scientific knowledge, usually by creating "certainty and strong cognitive order" from what is still contingent and controversial.[15] In other words, science popularization is not just a process of disseminating knowledge, but is itself part of knowledge production.

By foregrounding the disunity of science, these works call into question the image of science associated with rationality, objectivity, and universal truth, which is based on the assumption of a unified modern "science" that since the mid-nineteenth century had been translated from the West and fundamentally transformed the scientific and technological landscape in the Chinese society.[16] More importantly, these translations and transformations have unpacked a series of fundamental questions concerning "science" and the history of science: what constitutes "science" and hence a legitimate history of science; where, how, and for what purpose is "science" practiced; who counts as the proper practitioner and authority, and what are the standards of "science"? These questions urge us to discuss science as historicized and situated knowledge while approaching the production and circulation of scientific knowledge as social processes.

Recent scholarship in the history of science of the Mao era has joined the discussion by taking into account the multifariousness of the notion of "science" and its relationship to political leadership, scientists, low-level scientific workers, and the lay public in the second half of the twentieth century.[17] The edited volume *Mr. Science and Chairman Mao's Cultural Revolution* challenges the long-held view that science was destroyed in the Cultural Revolution.[18] Sigrid Schmalzer points out that the revolutionary leaders in red China embraced the same modernizing methods and economic and ideological goals like those of the Green Revolution conceptualized by the U.S. government in the Cold War era, that is, to "increase production and raise standards of living" by using mechanization, introducing new seeds, and applying modern chemicals. She also raised the vexed question whether the dichotomy of indigenous (*tu* 土) and foreign (*yang* 洋) science and experts, which the communist leadership integrated into its governing ideology in Maoist China, "decolonized" modern science or actually reproduced the terms of colonial epistemology.[19] Miriam Gross explores the role of grassroots science in the successful campaign against the snail fever in Maoist China and in consolidating the CCP's control in rural areas, calling for "a reassessment of the relationship between science and political control" during this period.[20] With a case study of China's prestigious Tsinghua University, Joel Andreas shows how the social concerns of reshuffling class identity and social mobility in educational institutions in the Mao era nevertheless ended up forging a new caste of technocratic officials that came to power in the post-Mao China. The CCP's failure to eliminate the distinction between mental and manual labor and to achieve their class-leveling ideal in the long run reveals the tight and intricate intertwinement of science with politics and society.[21]

The present study participates in rethinking science in Mao-era China by tracing the CCP's visions of science and its scientific thinking—both indispensable to the ruling party's conceptualizations of modernity—as well as their political, social, and cultural implications in Chinese society. The source materials consist of two complementary groups of data: first, materials of socialist mass culture—newspapers, magazines, films, posters, etc.—that convey the idea of science for, and sometimes of, the general audience; and second, materials promoting scientific and technological applications in industry and agriculture—manuals of and reports on new technologies and techniques, individual narratives of "advanced experiences," etc. Using published and publicly circulated sources does not mean we fully trust or subscribe to the contents of state propaganda materials, but rather that we take them seriously so as to understand how they functioned (or not) in different periods of the Mao era.[22] By analyzing these data we hope to present a complex and nuanced picture showing, first, how ideas of scientific knowledge, its producers, and methods were defined and redefined at different yet interconnected levels in the Mao era and second, how political and social concerns—such as national independence, socialism, the CCP's own ruling legitimacy and their social concerns of flattening hierarchy and allowing social mobility, etc.— were integrated into the theories and practices of making and circulating scientific knowledge, fluctuating with and reinforcing the truthfulness of "science."

While Tong Lam identifies a "passion for facts" in Republican China[23] that continued into the PRC era, the veracity of published sources in Maoist China is notoriously difficult to assess, if not totally unreliable. The numbers and statistic figures dating from the Great Leap Forward era and the second half of the Cultural Revolution (especially after the death of Lin Biao 林彪 in 1971), for example, claimed rapid economic growth and the improvement of living conditions. Yet as Yang Jisheng has argued in his detailed study on the famine of the Great Leap Forward, many statistical figures had little to do with the facts on the ground.[24] Nevertheless the obsession of using numbers as evidence shows a presumption that equalizes numbers with facts.

Arunabh Ghosh's 2020 study on statistics in Maoist China suggests, however, that one should not shy away from taking the discussions of statistics seriously and that dismissing Maoist mass science simply as antiprofessional and antiexpert can be unproductive.[25] Instead he traces the PRC's efforts of establishing a centralized statistical system in the early 1950s to solve the government's "crisis of counting." Statistics was considered as a technology of governance in the state-building process and its accuracy and completeness were the prime virtues. Yet, Ghosh shows, methods adopted at different periods varied so much that they ended up collecting different data, which were further processed by different ways of calculation and then produced very different figures. Ghosh's work thus helps us to understand how and why these statistical figures, including the unreliable ones, were made. His analysis includes many unpublished sources, such as letters, interviews, oral histories, and institutional archives, which are becoming increasingly difficult to access in recent years.[26] Workers and peasants who participated in science dissemination, in our case, usually leave very few written records. By looking at published sources, we focus on the issues about the methods and capacity of the state to formulate and disseminate a unified vision of science and technology across all levels of society.

What this study does NOT intend to offer is a "revisionist" argument to overwrite the human tragedies of famine, persecution, and death caused by political coercion in Maoist China, nor do we want to develop a Chinese theory of science and technology, or define science as an exclusively political act. At the risk of repeating ourselves, we want to emphasize that we do not subscribe to the dichotomy of science versus ideology; neither do we treat the terms "science" and "superstition" or "pseudo-science" as self-evident. The goal is to examine the making and dissemination of scientific knowledge as complicated and multifarious social processes that were negotiated through political and social concerns, limited resources, and not the least, various knowledge systems. The CCP's class politics, for example, foregrounded peasants and workers as active agents participating in the making and circulation of knowledge. Their needs, perspectives, as well as their (presumably) experience-based knowledge should contribute to what counted as scientific knowledge and how to disseminate it. The state's interpretation of Marxist-Leninist-Maoist ideologies managed to justify and incorporate the need of finding pragmatic ways to handle constrained financial, medical, and human resources on its way to modernization, which in turn led to the acknowledgment of pluralistic knowledge systems. We are going to show that "science" in the Mao era did not just receive knowledge of modern science, but also actively incorporated and recontextualized local knowledges. This, furthermore, means that the public opinions of science were far from uniform, which one may infer from the different claims to "scientific knowledge" in post-Mao China, which is supposed to be an era not only politically more relaxed but also more "scientific."

EXPERIMENT WITHOUT CONTINGENCY: THE PUBLIC DISCOURSE OF SCIENCE IN THE MAO ERA

Paradoxically, the epistemological complexity underlying the disunity of science in the pre-1978 era did not shake the authority of "science" as truthful and universal knowledge bringing about technological and social progresses. The reason lies in the fact that the notion of "science" has been part of the materialistic foundation of Marxism-Leninism and their Chinese interpretations, which, by valorizing the "natural" claim of science to objectivity and truthfulness, have consolidated their own discursive authority. Furthermore, experiment was established as a scientific method that would invariably succeed in the end and indeed found its new sites of knowledge production among the laboring people at their workplaces. The chance of failure in experiment, it was believed, would be tamed by choosing proper typical cases (*dianxing* 典型) and the experimenter's reflective use of his/her first-hand, empirical experience in the experiment. Experiments were carried out to test out new things in virtually all realms of the Chinese society, ranging from agricultural and industrial technologies and techniques to government policy-making and cultural production.

The elevation of "experiment" to the central position in the public discourse of science as *the* correct and non-contingent method of producing scientific knowledge is closely intertwined with Mao Zedong's own empirical approach to knowledge (*zhi* 知) by means of practice (*xing* 行), which he elaborated using the vocabulary of Marxist dialectical materialism in his long essay "On Practice" (*Shijian lun* 实践论), written in 1937 and published in full in the *People's Daily* (*Renmin Ribao* 人民日报) on December 29, 1950. Despite its historical and political specificities, Mao's essay deals with the classical debate in the history of science on the relationship between theory and practice, the philosophical/political ideal, and the application of scientific practices. According to Mao, one's social practices include activities in economic production, class struggle, political life, and activities in the arenas of science and art. Human beings discover truth of knowledge through their social practices and use practice to verify and develop the obtained knowledge, and thereby move human society and history to progress. The procedures of "practice, knowledge" and then "again practice and again knowledge" repeat themselves to bring the content of practice and knowledge to constantly higher levels.[27] His epistemology of practice, whose vocabulary of dialectic materialism legitimized its status as part of Marxism-Leninism, was disseminated through state propaganda apparatuses. With Mao's increasing authority, it had exerted powerful influence on the public discourse of science, to the extent that "On Practice" became the central text of his science philosophy.

On February 14, 1958, the Central Committee of the CCP asked its functionaries to farm, literally, "experiment plots" (*shiyan tian* 试验田) as a way to overcome bureaucratism. With this practice, the party cadres should integrate themselves into the masses, not only bettering their leadership but also turning themselves into red specialists, acquiring their knowledge of agricultural production through the practice of farming their plots.[28] The *People's Daily*'s editorial on the second day called for "farming experimental plots" in other economic sectors, crystalizing "experiment" as a scientific, empirical approach to true knowledge.[29]

An 1961 essay in the *People's Daily* cited Karl Marx to acclaim Francis Bacon (1561–1626) as "the real progenitor of English materialism and all modern experimental science."[30] Focusing on the dictum *"scientia potentia est"* (knowledge is power), this essay interpreted Bacon's idea of empirical science using Mao's vocabulary on practice: "Science means to use the rational methodology (*lixing fangfa* 理性方法) to sort out perceptual materials (*ganxing cailiao* 感性材料). Induction, analysis, comparison, observation, and experimentation are key means of the rational method." Bacon was also praised for his application of science to industry, promotion of technological inventions, emphasis on experience and experiment in understanding the world, as well as his advocacy of using scientific knowledge to control Nature.[31] Thus Bacon's ideas of science as portrayed in this essay endorse the officially sanctioned epistemology and application of science in China.

On February 16, 1963, an essay in the *People's Daily* claimed that "practice is the Marxist epistemology."[32] An editorial appearing in the newspaper on July 25, 1963—probably written by Mao himself—called for studying Marxist epistemology. It reiterated Mao's ideas in "On Practice" and highlighted "struggle of [economic] production, class struggle, and scientific experiment (*kexue shiyan* 科学试验)" as three most important social practices from which man's correct ideas originated. In the end of the editorial, they were hailed as "three great revolutionary movements to build a strong socialist country."[33] It is notable that here scientific experiment was perceived no longer just as a method to acquire and then verify knowledge in science and agricultural and industrial sectors, but also as a social practice (revolutionary movement) that should bring about social and political changes.

In September 1963, a *People's Daily* editorial continued to expand the social aspect of experiment by claiming that "experimenting with the typical case is a scientific method" in the process of policymaking. Experiment should be first carried out in a small scale in a work unit with representative conditions. The result would then be tested out and enriched in further experiments. The advantage of experimenting with the typical case, the editorial argued, was that it could prevent serious failure through learning the lesson from minor failures that may happen in the experiment. It thus resorted to the motto "failure is the mother of success" to interpret the meaning of failure positively, as offering valuable experiences.[34] Arunabh Ghosh's study of statistics and statecraft in the early PRC notes that choosing typical case derived from Mao Zedong's social investigation method in his early years to research on the peasant movement in Hunan, which Ghosh terms the ethnographic approach to social reality. Built on Mao's epistemology of practice, this method assumes that a "'typical' ... understanding could be extrapolated to produce wider, more comprehensive knowledge of social, economic, or cultural trends."[35] This method had not only impacted PRC statistic work and its drastic methodological change in 1958,[36] but seems also to underlie the discourse of experiment in the policymaking process in September 1963. As a method, the approach of working with the typical case places much weight on the quality of the experimenter who, one would expect, would develop effective criteria to identify the typical case, to collect first-hand data (observing, recording, interacting with the objects), and to analyze the data. But as Ghosh argues, Mao's social investigation method generally "valorized the personal experience" of the investigator while "discounted methodological issues such as representativeness and comprehensiveness."[37]

The experiments of scientific farming, as another editorial in the *People's Daily* in 1965 argued, should, instead of starting from a theoretical hypothesis, be rooted in the practical needs to test out and verify peasants' empirical experiences: problems in real economic production should be identified, peasants' experiences should be taken as the departure point of experiment, and results from experiment field, laboratories, and demonstration farms should all be taken into consideration.[38] Similar to the process of policymaking, the research results of scientific farming should be promoted, tested out, developed, and supplemented by further experiments that bring together the farming experiences of both the peasant and the low-level scientific worker on the ground: this procedure was summarized as experiment, demonstration, and extension in the agricultural scientific experiment movement.[39]

Ironically, the personal cult of Mao Zedong in the Cultural Revolution reached such a level that his dialectics of practice had ossified into a dogma that was applicable on everything, ranging from guiding spring farming and raising pigs scientifically to revolutionizing families.[40] A 1975 volume interpreting Mao's "On Practice" and his influential 1963 essay "Where Do Correct Ideas Come From?" (*Ren de zhengque sixiang shi cong nali laide?* 人的正确思想是从哪里来的?), for example, argued not only that practice functions to measure "thoughts, theories, policies, plans, and methods" but also that "correct thoughts" have the power to produce material results, a view closer to idealism than materialism: "Once the correct thoughts representing the progressive class are grasped by the masses, they will turn into material force to reform the society and the world."[41]

Schmalzer observes that the parable of "The Foolish Old Man Moves the Mountain (*Yugong yishan* 愚公移山)," which teaches the moral of persistence and perseverance, was constantly used to characterize scientific experiments conducted by sent-down youths in 1960s and the 1970s rural China, but it did not necessarily match the youth's true experiences.[42] The parable and motto sought to establish the faith and optimism that scientific experiment was bound to succeed: by foreclosing the possibility of the unknown and failure, they suggest that the prospect of experiment is predictably positive.[43] The public discourse of scientific experiment moved from a scientific method of knowledge acquisition, economic production, policymaking to a revolutionary movement in its own right. Given that the term "revolution" had firmly established its discursive authority as a social practice with teleologically progressive outcomes by 1963, "scientific experiment" served to consolidate the positive prospect of the larger, equally non-contingent "experiment" of socialist modernity.

Despite the oscillation of the state's policy about knowledge production and dissemination and the constant tension and confusion they caused, the authority of "science" was never undermined, even when the official historiography in the PRC tended to portray Deng Xiaoping's 邓小平 (1904–97) call to modernization in 1978 as a farewell to revolution and as a radical rupture from the Maoist ideal of social practice. Pragmatism and professional science have replaced ideological craze, voluntarism, and class struggle in post-Mao China—that is the view that has been accepted by many historians until last decades.[44] Taking 1978 as a historical moment of rupture in science policy and practice, however, runs the risk of underestimating the scientific breakthroughs in the Mao era while overemphasizing the universalist character of science. Viewing both the reform and the opening to the outside world as fundamental factors to restore science's autonomy after the Cultural Revolution may also turn out to endorse the CCP's own discourse about its progressive role in ruling China, which dismissed the Great Leap Forward campaign and the Cultural Revolution as deviating, abnormal periods on its way to the historically destined leadership.

Though scientific experiment as *the* primary approach to knowledge and various scientific miracles deployed from the 1950s to the 1970s were overruled after 1978, they had paradoxically prepared, in an accumulative way, for the ready acceptance of (the power of) professional—or other forms of—science in the reform era. The fact that the catchphrase in post-Mao China—the Four Modernizations (*sige xiandaihua* 四个现代化)—was first phrased and declared in 1964, right in the middle of the Mao era, bespeaks the legacy of the Mao era to the years after it. In fact, the strong belief in science and technology fostered and nurtured in the revolutionary era can be considered preparations for today's China that, according to James Wilson and James Keeley, is on the way to becoming the next scientific powerhouse.[45] This seems plausible in view of China's constantly growing investment in research, large engineering projects, and infrastructure construction, such as the Three Gorges Dam and the high-speed railway network. Despite controversies, they allow the party-state to prove itself to be the effective, successful modernizer of the nation and therefore *the* legitimate ruling Party with an ideology of developmentalism based on science and technology.

NOTES

1. Mao Zuoben 茅左本, *Women zuxian de chuangzao faming* 我们祖先的创造发明 (Inventions of Our Ancestors) (Shanghai: Laodong chubanshe, 1951).
2. Mao, *Women zuxian de chuangzao faming* (1951), 101–2.
3. Mao, *Women zuxian de chuangzao faming* (1951), 9.
4. Mao, *Women zuxian de chuangzao faming* (1951), 100.

5. Mao Zuoben 茅左本, *Women zuxian de chuangzao faming* 我们祖先的创造发明 (Inventions of Our Ancestors) (Shanghai: Renmin chubanshe, 1957), 138.

6. Mao, *Women zuxian de chuangzao faming* (1957), 129.

7. When founded in December 1955, Prime Minister Zhou Enlai contributed the following words of encouragement: "Developing the medical and pharmacological heritage of our nation and serving the socialist reconstruction (发展祖国医药遗产，为社会主义建设服务)," see *Zhou Enlai dacidian* 周恩来大辞典 (Great Dictionary on Zhou Enlai), ed. Cao Guangzhe 曹光哲, Qi Pengfei 齐鹏飞, and Wang Jin 王进 (Guilin: Guangxi chubanshe, 1997), 1185. In 2005, the academy was renamed China Academy of Chinese Medical Sciences 中国中医科学院 on the occasion of its fiftieth anniversary.

8. Sigrid Schmalzer, *Red Revolution, Green Revolution: Scientific Farming in Socialist China* (Chicago: University of Chicago Press, 2016), 14.

9. For an example of such propaganda, see Du Runsheng, "Great Progress Made in the Natural Sciences in China During the Last Decade," *The Science News-Letter* 78, no. 24 (1960): 377–92, originally appearing in *Scientia Sinica* VIII, no. 11 (1959).

10. Examples for how political interventions hinder scientific progress are, for instance, the Lysenko affair (Laurence Schneider, *Biology and Revolution in Twentieth-century China* [Lanham: Rowman & Littlefield, 2003]), the campaign against Einstein's general theory of relativity (Danian Hu, *China and Albert Einstein: The reception of the Physicist and His Theory in China, 1917–1979* [Cambridge, MA: Harvard University Press, 2005]), or the willful persecution of Shu Xingbei 束星北 who even as a mathematician was not spared (Huang Yong 黄勇, "Youpai wenxue zhong de ziran kexuejia 右派文學中的自然科學家 (A Natural Scientist in the Rightist Literature)," *Ershiyi shiji*, no. 110 [2008], 79–88). A more positive attitude toward the scientific achievements in the Mao era can be found in *Mr. Science and Chairman Mao's Cultural Revolution: Science and Technology in Modern China*, ed. Chunjuan Nancy Wei and Darryl E. Brock (Lanham: Lexington Books, 2013); and Joshua Eisenman, *Red China's Green Revolution: Technological Innovation, Institutional Change, and Economic Development under the Commune* (New York: Columbia University Press, 2018).

11. Fa-ti Fan, "East Asian STS: Fox or Hedgehog?" *East Asian Science, Technology and Society* 1, no. 2 (2007): 243–47; Warwick Anderson, "Postcolonial Specters of STS." *East Asian Science, Technology and Society* 11, no. 2 (2017): 229–33.

12. Ian Hacking, *Representing and Intervening: Introductory Topics in the Philosophy of Natural Science* (Cambridge: Cambridge University Press, 1983); Peter Galison and David J. Stump, eds., *The Disunity of Science: Boundaries, Contexts, and Power* (Stanford: Stanford University Press, 1996); Sandra Harding, *Sciences from Below—Feminisms, Postcolonialities, and Modernities* (Durham and London: Duke University Press, 2008).

13. Hacking, *Representing and Intervening*, 149–54.

14. Karin Knorr-Cetina, "The Care of the Self and Blind Variation: The Disunity of Two Leading Sciences," in *The Disunity of Science: Boundaries, Contexts, and Power*, eds., Peter Galison and David J. Stump (Stanford: Stanford University Press, 1996), 287–310.

15. Terry Shinn and Richard Whitley, eds., *Expository Science: Forms and Functions of Popularisation* (Dordrecht: Lancaster, 1985), 23.

16. Daniel Kwok, *Scientism in Chinese Thought 1900–1950* (New Haven: Yale University Press, 1965); Thomas Fröhlich, *Staatsdenken im China der Republikzeit (1912–1949): Die Instrumentalisierung philosophischer Ideen bei chinesischen Intellektuellen* (Frankfurt: Campus Verlag, 2000); Lydia Liu, *Translingual Practice: Literature, National Culture, and Translated Modernity—China, 1900–1937* (Stanford: Stanford University Press, 1995); Xiong Yuezhi 熊月之, *Xixue dongjian yu wan Qing shehui* 西学东渐与晚清社会 (Western Learning Spreading to the East and Late Qing Society) (Shanghai: Renmin chubanshe, 1994).

17. In a different context E. Perry has argued that neither the diversity nor the legacy of the Mao era should be overlooked. See Elizabeth J. Perry, "The Promise of PRC History," *Journal of Modern Chinese History* 10, no. 1 (2016): 113–17.

18. Wei and Brock, *Mr. Science and Chairman Mao's Cultural Revolution*.

19. Schmalzer, *Red Revolution, Green Revolution*, 2–3, 34–38.

20. Miriam Gross, *Farewell to the God of Plague: Chairman Mao's Campaign to Deworm China*. (Berkeley: University of California Press, 2016), 9–12.

21. Joel Andreas, *Rise of the Red Engineers: The Cultural Revolution and the Origin of China's New Class* (Stanford: Stanford University Press, 2009).

22. Alternatively, one could also consider archival sources, diaries, and memoirs, yet such sources are not unproblematic either, as Sigrid Schmalzer argues in, "Youth and the 'Great Revolutionary Movement' of Scientific Experiment in 1960s–1970s Rural China," in *Maoism at the Grassroots: Everyday Life in China's Era of High Socialism*, eds., Jeremy Brown and Matthew D. Johnson (Cambridge, MA: Harvard University Press, 2015), 154–78.

23. Tong Lam, *A Passion for Facts: Social Surveys and the Construction of the Chinese Nation-State, 1900–1949* (Berkeley: University of California Press, 2011).

24. Yang Jisheng, *Tombstone: The Untold Story of Mao's Great Famine* (London: Allen Lane, 2012).

25. Arunabh Ghosh, *Making it Count: Statistics and Statecraft in the Early People's Republic of China* (Princeton: Princeton University Press, 2020), especially chapter 8, "A 'Great Leap' in Statistics."

26. The access has radically decreased in recent years. See Sören Urbansky and Arunabh Ghosh, "Introduction," *The PRC History Review* 2, no. 3 (June 2017): 1–3.

27. Mao Zedong, "On Practice," in *Selected Works of Mao Tse-tung*, Vol. I (Beijing: Foreign Languages Press, 1977), 295–309.

28. "Zhonggong zhongyang guanyu pubian tuixing zhong shiyantian de jingyan de tongzhi 中共中央关于普遍推行种试验田的经验的通知 (A notification of the Central Committee of the Chinese Communist Party concerning the experience of promoting experimental plots)," in Zhonggong zhongyang wenxian yanjiushi 中共中央文献研究室, *Jianguo yilai zhongyao wenxian xuanbian* 建国以来重要文献选编, Vol. 11 (Beijing: Zhongyang wenxian chubanshe, 1992–2011), 150–151.

29. Editorial, "Zhong shiyantian—you hong you zhuan de daolu 种试验田—又红又专的道路 (Farming experimental plots—the path to the red expert)," *Renmin Ribao*, February 15, 1958; Editorial, "Gongye zhong jianli 'shiyantian' de xin jingyan

工业中建立'试验田'的新经验 (New Experiences on the Establishment of Experimental Plots in the Industry)," *Renmin Ribao*, March 24, 1958.

30. Karl Marx and Friedrich Engels, "Critical Battle against French Materialism," in *The Holy Family or Critique of Critical Criticism Against Bruno Bauer and Company* (chapter VI 3.d., 1845), Marx/Engels Archive, last accessed December 11, 2018, https://www.marxists.org/archive/marx/works/1845/holy-family/ch06_3_d.htm. Miao Litian 苗力田, "'Zhishi jiushi liliang'—jinian Fulanxisi Peigen dansheng sibai zhounian '知识就是力量'—纪念弗兰西斯．培根诞生四百周年 (Knowledge is Power—Commemorating the 400th birthday of Francis Bacon)," *Renmin Ribao*, January 22, 1961. The author cited from the Chinese translation (1957) of Karl Marx and Friedrich Engels, "Critical Battle against French Materialism" in their The Holy Family or Critique of Critical Criticism against Bruno Bauer and Company (chapter VI 3.d., 1845). See https://www.marxists.org/archive/marx/works/1845/holy-family/ch06_3_d.htm (last accessed August 8, 2021).

31. Ruoshui 若水, "Makesi zhuyi de renshilun shi shijianlun 马克思主义的认识论是实践论 (The Epistemology of Marxism Is the Text *On Practice*)," *Renmin Ribao*, February 16, 1963.

32. Editorial, "Renzhen xuexi Makesi zhuyi de renshilun 认真学习马克思主义的认识论 (Earnestly Learning the Epistemology of Marxism)," *Renmin Ribao*, July 25, 1963.

33. Editorial, "Dianxing shiyan shi yige kexue de fangfa 典型试验是一个科学的方法 (Experimenting with the typical case is a scientific method)," *Renmin Ribao*, September 20, 1963.

34. Ghosh, *Making it Count*, 5.

35. Ghosh, *Making it Count*, introduction, chapter 2, 47–53, and chapter 8.

36. Ghosh, *Making it Count*, 48.

37. Editorial, "Banhao sanjiehe de yangbantian, cujin nongye kexue shiyan yundong 办好三结合样板田，促进农业科学实验运动 (Managing well the three-times integrated model field, promoting the scientific experiment movement in agriculture)," *Renmin Ribao*, March 28, 1965.

38. Schmalzer, *Red Revolution, Green Revolution*, 40.

39. Hebei baxian luoboying dadui dangzhibu 河北霸县罗卜营大队党支部, "Yong bianzhengfa zhidao chungeng shengchan 用辩证法指导春耕生产 (Using dialectics to guide spring plowing and production)," *Renmin Ribao*, March 25, 1970; Cao Xinhua 曹新华, "Yunyong weiwu bianzhengfa, gaohao jiating geminghua 运用唯物辩证法，搞好家庭革命化 (Using materialistic dialectics to revolutionize the family)," *Renmin Ribao*, October 6, 1970; Duan Yukun 段玉坤, "Xuexi weiwu bianzhengfa, shixian yangzhu kexuehua 学习唯物辩证法，实现养猪科学化 (Studying materialistic dialectics to implement the scientific way of raising pigs)," *Renmin Ribao*, September 5, 1971.

40. Shi Zhe 施哲, "*Shijianlun*" he "*Ren de zhengque sixiang shi cong nali laide*" qianshuo《实践论》和《人的正确思想是从那里来的》浅说 (Changchun: Jilin renmin chubanshe, 1975), 17. On the 1963 essay see Mao Zedong 毛泽东, "Ren de zhengque sixiang shi cong nali laide 人的正确思想是从哪里来的？(1963)," in

Mao Zedong zhuzuo xuandu (xiace) 毛泽东著作选读（下册） (Beijing: Renmin chubanshe, 1986), 839–41.

41. Schmalzer, "Youth and the 'Great Revolutionary Movement' of Scientific Experiment in 1960s–1970s Rural China,."

42. On the impossibility of accepting the unknown in the Maoist science discourse see Marc Matten, "Coping with Invisible Threats: Nuclear Radiation and Science Dissemination in Maoist China," *East Asian Science, Technology and Society* 12, no. 3 (2018): 235–56.

43. Works in the history of science emphasizing a break with the Mao era appeared mostly in the early reform era, see for example, Erik Baark, "The Structure of Technological Information Dissemination in China: Publication of Scientific and Technological Manuals, 1970–77," *China Quarterly*, no. 83 (September 1980): 510–34; Kieran P. Broadbent, *Dissemination of Scientific Information in the People's Republic of China* (Ottawa: International Development Research Centre, 1980); Hong Yung Lee, *From Revolutionary Cadres to Party Technocrats in Socialist China* (Berkeley: University of California Press, 1991); Leo A. Orleans, ed., *Science in Contemporary China.* (Stanford: Stanford University Press, 1980).

44. James Wilson and James Keeley, *China, the Next Science Superpower? The Atlas of Ideas: Mapping the New Geography of Science* (London: Demos, 2007). See here also the recent work by Robert A. Rhoads, Xiaoyang Wang, Xiaoguang Shi, and Yongcai Chang, eds., *China's Rising Research Universities: A New Era of Global Ambition* (Baltimore: Johns Hopkins University Press, 2014).

Chapter One

Defining Correct Science
Transformations of Knowledge Epistemologies

In 1932, some political and intellectual elites of Republican China founded the Association for the China Scientization Movement (*Zhongguo kexuehua yundong xiehui* 中国科学化运动协会) and launched its magazine *Scientific China* (*Kexue de Zhongguo* 科学的中国) in the following year.[1] The chief-editor Zhang Qiyun 张其昀 (1901–85), a renowned geographer and historian, stated its goals in the inaugural issue as follows:

> We gathered many natural scientists and researchers of applied science here in order to bring scientific knowledge to the ordinary people (*minjian* 民间) so that it will become common wisdom. We further hope that after this knowledge is disseminated among the people, it will produce dynamics to vitalize our nation, which has come to its life-and-death moment, and to rejuvenate our Chinese culture, which has been declining with each passing day.[2]

What, then, is a "scientific China" envisioned by these elites in the 1930s? While the aim of (disseminating knowledge of) science was largely undisputed in the history of twentieth-century China, its definition and its epistemologies were not. Kwok has shown in his monograph *Scientism in Chinese Thought 1900–1950* that the intellectual elite in the Republican era considered science and technology as fundamental to the survival of China as a nation-state. Facing the dominance of brutal social Darwinism and Western imperialism on a global scale, the Chinese elite, whose majority had studied in Japan, Europe, and the United States, widely accepted the superiority of Western modern science. The quote of Zhang Qiyun demonstrates that the dissemination of science and the implementation of scientific technologies, widely regarded as patriotic duties of the Republican elites, were believed to be achieved in a top-down way, in which the state and the foreign-educated scientist worked together to determine the legitimate forms of "scientific" knowledge. Their ambitions to establish modern science in Chinese society were accompanied by efforts of eradicating other forms of knowledge that were castigated as superstitious (*mixin* 迷信). The discourse of a hierarchically organized and apparently universal science, however, faced from time to time an uphill struggle when encountering local and pre-modern knowledges based on different epistemologies.

The constant conflicts and negotiations between foreign and indigenous or/and traditional and modern knowledges have largely shaped the discourse of science, as shown by the acclaimed publications of, among others, Nathan Sivin, Chu Pingyi, Rebecca Nedostup, and Hsiang-lin Lei.[3] They did not end after the founding of the People's Republic. While the PRC inherited many aspects of the discourse of science from the May Fourth era and adapted them to the tenets of socialism, this by no means led to a unified and undisputed notion of science.[4] The epistemological transformation to modern science had been impacted by translation that included the acceptance of exactitude and factual knowledge as central aspects of science at the turn of the twentieth century, as Tong Lam discusses in his book *A Passion for Facts* (2011). When Hu Shi 胡适 (1891–1962), a U.S.-educated scholar and one of the most influential cultural leaders of the May Fourth era, published his satirical *Biography of Mister More-or-Less* (*Chabuduo xiansheng zhuan* 差不多先生传) in 1924, he criticized inexactitude as a rampant and most despicable trait among his fellow countrymen. In the preface of the book *Science and the View of Life* (Kexue yu renshengguan 科学与人生观) Hu was even more blunt when lamenting that "the Chinese view of life has never even encountered science face-to-face! At this time, we are still troubled that science is not being promoted adequately, troubled that science education is not being developed, and troubled that the force of science is not enough to sweep away the evil spirit that spreads all over the country."[5] The disregard of accuracy in measuring and in (clock) time were also a key issue in Arthur Smith's (1845–1932) analysis of *Chinese Characteristics* in 1894 when he wrote that "the existence of a double standard of any kind, which is often so keen an annoyance to an Occidental, is an equally keen joy to the Chinese. Two kinds of cash, two kinds of weights, two kinds of measures, these seem to him natural and normal, and by no means open to objections." In his words, "in China a 'pint' is not a pint, nor is a 'pound' a pound."[6] This was extremely irritating for the American missionary, because all standards were relative and thus could not be subjected to a universal rationality as constants. Whereas Smith held the opinion that a theoretically sound science in the modern sense could hardly evolve in China, he had forgotten that the universality of both clock time and modern science was also a new thing for his own compatriots. His contempt of inexactitude in Chinese daily life thus exuded a sense of superiority originating from Eurocentrism.[7]

A few years after Hu Shi's scorching criticism, a writer under the pseudonym Chaosheng 潮声 took up this view in 1931 when discussing the need of *seeking truth from facts* (*shishi qiushi* 实事求是). He argued that the wanton (*suisui bianbian* 随随便便) and sloppy (*mama huhu* 马马虎虎) attitudes of the Chinese people would be of no help in advancing the nation. The need of exact measuring was closely linked to factual knowledge, whose effort of quantification shows the rationalization of results from observation and experimentation.[8] The passion for facts that Tong Lam identified for Republican China with help of Chaosheng's text would continue into the PRC era when the slogan *shishi qiushi* was elevated to the central tenet of communist political philosophy by Mao Zedong who presented the four characters as an inscription to the Central Party School in Yan'an in 1941. Consistent with his emphasis on "practice," Mao meant to remind the cadre of getting hold of actual problems instead of just engaging in idle talk (*shishi qiushi, bushang kongtan* 实事求是、不尚空谈). It was the spelled-out intention of the Party to emphasize its ability to solve problems and to improve the living standards of the population, instead of engaging in—what was judged by them as bourgeois—idleness grounded in metaphysics and idealism.[9]

Before spelling out how Mao Zedong's explicit and repeated advocacy of practice-oriented, fact-based epistemology of knowledge impacted the acceptance of scientific socialism as a way to understand and transform both human society and the natural world, we describe in the following part how the disunity of science was created in the era of socialist construction after 1949. Taking the well-studied case of human medicine, we show that the dichotomy of science and superstition was not so stark and that, in addition to the findings of Benjamin Elman in his 2005 work on science in China, there was more than Chinese medicine that "survived" in the wake of Western impact. We do so by avoiding the question whether or when some knowledge can count as scientific and by taking Chinese knowledges seriously as local knowledges, instead of viewing them as alternative to or derivate from Western scientific knowledge.

HEALING SCIENCE

In order to protect and strengthen the nation that since the late imperial era had been disparaged as the Sick Man of East Asia,[10] medical professionals that had been educated abroad since the 1900s introduced new ways of understanding medical processes, diagnosis, and treatment of ailments. This modern knowledge followed the principles of positivist science[11] and was seen in contrast to older medical knowledge, or Chinese medicine (henceforth CM) that now came under attack.[12] In place of the holistic concept of medicine—which was only termed Traditional Chinese Medicine (TCM) in the recent past—the evidence-based form of medicine gained ground in the twentieth century.[13] Describing diagnosis and evaluating the success of treatment through quantitative measures, it has come to be known as modern biomedicine. These two concepts of medicine are believed to differ fundamentally in their respective approach to the understanding of sickness and health.

The introduction of modern medicine in the late nineteenth century appeared convincing to the Chinese elite, because it seemed to be able to explain to them why the European nations and Japan were able to command stronger armies, both physically and psychologically. When Japan annexed Taiwan after its victory in the Sino-Japanese War of 1894–95, it pursued a sweeping policy of modernization that included the attempt of abolishing CM on the island. During the May Fourth Movement, intellectuals and political leaders such as Chen Duxiu 陈独秀 (1879–1942), Lu Xun 鲁迅 (1881–1936), Sun Yat-sen 孙逸仙 (1866–1925), and Hu Shi followed suit by calling for replacing Chinese medicine with Western/modern medicine.[14] As shown by Rebecca Nedostup, the Nationalist government under the leadership of the Kuomintang (KMT) expanded the medical debate to fights against superstition in the Nanjing decade (1927–37) by turning against all other forms of healing practitioners such as shamans and sorcerers, as well as geomancers, physiognomists, fortune tellers, and secret societies. It was part of its modernization agenda to provide both physical and psychological healings to the population.[15] *Mixin* (superstition) no longer referred to a matter of false religion (or *xiejiao* 邪教), but rather a matter of having any religion at all. Not impurity but a strict sense of positivist rationality defined the line between science and superstition,[16] and the fight against superstition could surely not exclude the medical field. For the physician Yu Yunxiu 余云岫 (1879–1954), who had studied medicine in Japan and become the first president of the Medical Practitioners' Association (*Shanghai yishi gonghui* 上海医师公会) founded in 1925 in Shanghai, the choice for the "right" medical system had to be implemented by political means and state power:

> I think that without the power of politics, there is no way to popularize scientific medicine in China. If we keep focusing on advertising [scientific medicine] to the mass, no one knows if there will be any effect at all in one hundred, or even one thousand, years.[17]

When the Republic established its Ministry of Health in Nanjing in 1928, it not only provided for the first time in Chinese history a national administrative center taking charge of improving the health of its citizens, that is to achieve "hygienic modernity,"[18] but also subjected medical knowledge to the control of state institutions. In early 1929, the first National Public Health Conference—dominated by Western medicine–trained physicians—submitted a proposal to "Abolish old-style medicine in order to clear away the obstacles to medicine and public health."[19] As shown by Hsiang-lin Lei, the reaction of Chinese medicine practitioners to this 1929 decision was to organize themselves into the National Medicine Movement (*Guoyi yundong* 国医运动), arguing that Chinese medicine was one measure to resist cultural imperialism. The controversy created a "bifurcated medical field" in China that is still present today.[20] There was thus not a synthesis (*jiehe* 结合) of Chinese and Western medicines, Lei argues, but—due to the politicization of the field—rather the idea that Chinese medicine was epistemologically different because it was based on *jingyan* 经验, or experience, instead of abstract, theoretical knowledge.

In medical discourse of that time, local forms of medicine were not only called old medicine (*jiuyi* 旧医), but more often than not referred disparagingly to as black magic (*wushu* 巫术, or *xieshu* 邪术) and thus put into the category of the unscientific (*feikexue* 非科学). This category describes a mode of thinking that is not in accordance with the notion of (modern) science, which itself is based on the assumption that there is a systematic way of building and organizing knowledge along the exclusive categories of false/correct. It aims at describing and explaining natural phenomena by logical and rational principles, instead of accepting handed-down wisdom as self-evident. The opposition of modern and Chinese medicines is, at best, normative, and Hsiang-lin Lei has meticulously shown how both medical systems evolved in relation to and interaction with each other. According to his reading, Chinese medicine was able to change its image of being antithetical to science and to challenge the universalist conception of modern/Western science.[21]

We intend to show in this book that the assumption of a clear-cut and unequivocal opposition of different conceptions of scientific practices is not unique in medical science. It can also be found in other fields, where the state pragmatically downplayed the confrontation between "Western science" and "Chinese pre-scientific knowledge" in its negotiation between the goal of modernization and constrained resources, be it agricultural mechanization, steel production, or veterinary medicine. In the Mao era, this was possible despite the dominance of Marxism-Leninism that aimed to make historical materialism and dialectics of nature the foundation of legitimate science. While taking the Soviet Union as its model, the Maoist discourse of science shared in principle the Nationalists' conviction that superstitious practices such as shamanism and exorcism hindered modernization. Yet, the insistence on experiment and the Maoist pragmatic epistemology of knowledge allowed a more heterogeneous understanding of scientific practices.

In November 1949, the *People's Daily* praised the efforts of Soviet experts to help the Chinese people to wipe out the bubonic plague in Inner Mongolia and Chahar Province. By emphasizing the need for hygiene and quarantine isolation for the sick, the Soviet medical experts propagated scientific medical knowledge and thus turned against superstitious practices that insisted on prayers and offerings to the gods.[22] In the following week, a group of more than one hundred doctors vaccinated 100,000 people, and school teachers started their propaganda work by making use of the local broadcasting stations, urging the population to kill rats, get rid of garbage, vaccinate people, and isolate sick persons. In the end, the health and living situation of the Chinese people reportedly improved significantly.[23]

At that time, the CCP understood its fight against superstition as a necessary step to ensure the economic well-being of the people, arguing that anti-scientific activities impeded agricultural production and resulted in a waste of resources, when people went on searching for holy water and neglected farming, or spent their money on buying incense or purchasing ritual items.[24] These early PRC official reports had a remote resonance with the late Qing government's efforts to contain the Manchurian Plague in 1910 and 1911, but they also demonstrated an internationalism at the beginning of the Cold War. The trust in modern medicine and the argument against folk religious healing practices as wasteful of human and material resources in these reports attest to a continuity in scientific thinking across the 1949 divide.

The need to subordinate medical science to economic reconstruction resulted in its institutionalization in the early years of the PRC. After the First and Second National Conferences on Public Health in 1950 and 1951 the CCP defined the "four great guiding principles" of health policy, that is 1) the task of medicine was to serve the workers and farmers, 2) prevention was more important than curing, 3) Chinese medicine and Western biomedicine should be united, and 4) the people's health had to be improved by making use of mass movements.[25] Partly due to the continuous influence of Yu Yunxiu—who had participated in the First National Conferences on Public Health in Beijing in 1950—Chinese medicine was still regarded a relic of the feudal era. Accordingly, when Chinese medicine practitioners obtained their license in 1952 via state examinations, they were also required to have knowledge of Western medicine.[26]

As argued by Sigrid Schmalzer in her monograph on the Peking Man, science dissemination in the 1950s meant to replace wrong beliefs with scientific knowledge. In other words, propagating the correct knowledge always went hand in hand with rejecting the false one. Instead of "seeking the gods and begging for medicine" (*qiushen taoyao* 求神讨药), the people must be enlightened properly. The CCP especially felt that the rural population needed such enlightenment, because they were believed to have relied for centuries on supernatural forces. Accordingly, the fight against epidemic diseases in rural areas received special attention in the state's media, which usually followed a three-step scheme that was typical for science dissemination activities in that era: first the superstitious phenomenon was refuted, then it was identified as an issue of lacking (political) education, and finally it was emphasized how scientific knowledge could contribute not only to the health of the individual but also to socialist construction and modernization. As an example, let us take a look at the following lantern slideshow materials dating from 1951, entitled *Preventing Contagious Diseases in the Summer Season* (*Yufang xiaji chuanranbing* 预防夏季传染病).²⁷

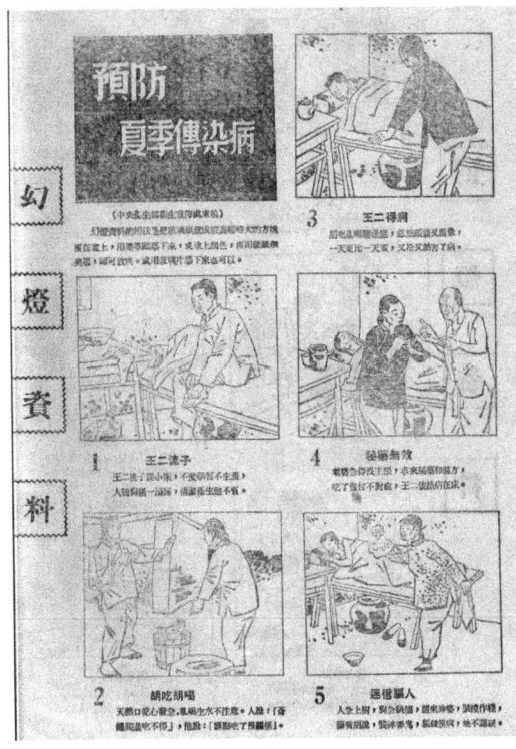

Figure 1.1. Zhongyang weishengbu weisheng xuanchuanchu 中央卫生部卫生宣传处, "Yufang xiaji chuanranbing 预防夏季传染病 (Preventing Contagious Diseases in the Summer Season)—Part I," *Kexue puji gongzuo*, no. 16 (June 1951): 179.

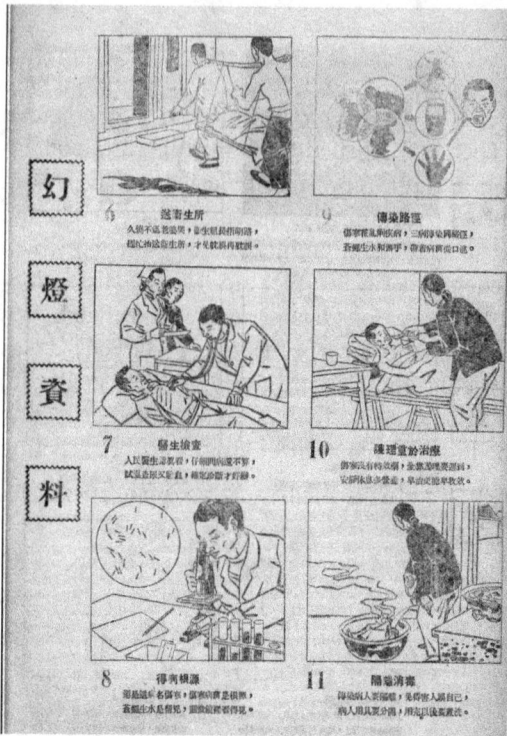

Figure 1.2. Zhongyang weishengbu weisheng xuanchuanchu 中央卫生部卫生宣传处, "Yufang xiaji chuanranbing 预防夏季传染病 (Preventing Contagious Diseases in the Summer Season)—Part II," *Kexue puji gongzuo*, no. 16 (June 1951): 180.

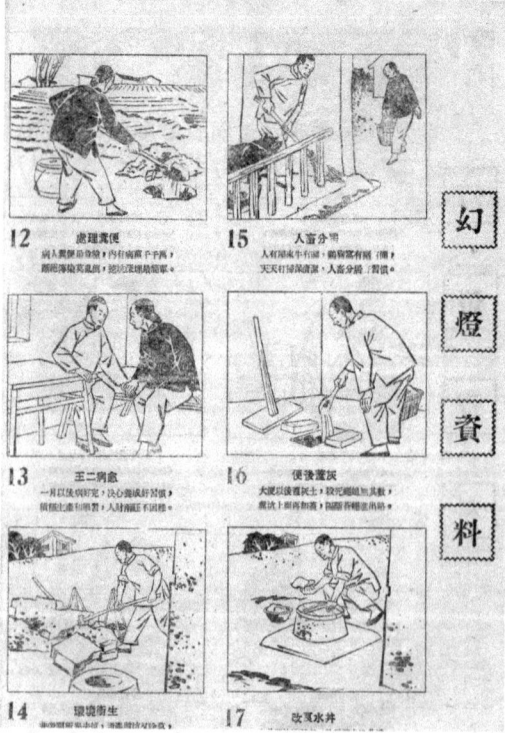

Figure 1.3. Zhongyang weishengbu weisheng xuanchuanchu 中央卫生部卫生宣传处, "Yufang xiaji chuanranbing 预防夏季传染病 (Preventing Contagious Diseases in the Summer Season)—Part III," *Kexue puji gongzuo*, no. 16 (June 1951): 181.

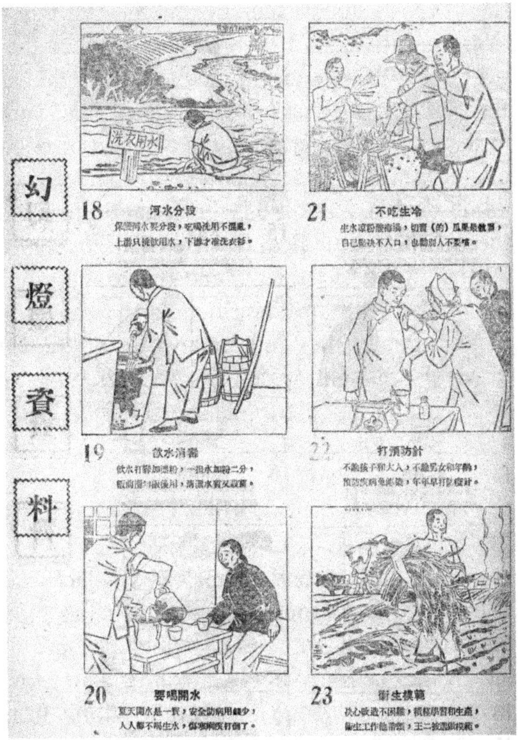

Figure 1.4. Zhongyang weishengbu weisheng xuanchuanchu 中央卫生部卫生宣传处, "Yufang xiaji chuanranbing 预防夏季传染病 (Preventing Contagious Diseases in the Summer Season)—Part IV," *Kexue puji gongzuo*, no. 16 (June 1951): 182.

The lantern slides show a peasant called Wang Er who is depicted as lazy and unproductive. He becomes sick due to his neglect of basic hygienic rules (slide 1). His wife turns to a local healer who prescribes ineffective herbs (slide 4), and even a spiritual medium called by her is unable to expel the evil spirits that are believed to be the cause of the illness (slide 5). Medical workers at a local health clinic—wearing the lab coat as a sign of professionalism—then identify bacteria as the major culprit (slide 8) and convince Wang Er of this fact by showing the pathogens under the microscope, the iconic instrument of modern medicine at that time.[28] In the end, Wang changes his lifestyle after recovery (slides 10–13), pays attention to personal hygiene, and becomes a productive member of society (slides 13–22). By discarding old habits and old belief/knowledge, Wang makes himself fit for contributing to socialist construction (slide 23).

The intention of the slide show lies in transforming society by teaching proper behavior, in this case the basic principles of hygiene. While the improvement of hygiene (*weisheng* 卫生) had already been an important concern in large coastal cities since the Republican era,[29] the communication of such knowledge proved more difficult on the countryside, which was due to the lower degree of literacy and the fact that terminology and knowledge of Western biomedicine still had not reached the rural population. This was the prime reason why visual media such as the lantern slides were the preferred form of science dissemination materials, similar to the propaganda posters at that time. Furthermore, the concern of being understood by the addressed audience was also a reason why the slide show used the term *shanghan* 伤寒 (slides 8–10), a traditional designation for cold damage disorders taken from the traditional medical treatise *Shanghan Lun* 伤寒论 (compiled by Zhang Zhongjing 张仲景 at the end of the Han dynasty, 202 BC–220 AD), even though bacteria are identified as the cause of sickness. Paying attention to local characteristics and frameworks of knowledge was an important element of Maoist thought, which has contributed to and accommodated the disunity of science in the science discourse of the PRC, as this book is going to show.

DEBATING MODERNITY

Since the early twentieth century, science—with its ability to create factual knowledge and productive citizens—has stood in the center of the public discourse about how China could become modern. While the KMT state envisioned in many fields modern science knowledge as irreconcilable with traditional knowledge—a view resulting from the acknowledgment of the West as being superior after the arrival of imperialism—the resilience and resistance of knowledge from the pre-modern era is seen today as an act of necessary postcolonial self-assertion that questions the universal character of Western modernity, as many contributions to *East Asian Science, Technology and Society* (*EASTS*) exemplify.[30] In the field of the history of science, the self-claimed purity of Western modernity and modern science has been put into question since Bruno Latour provocatively claimed in 1991 that *We Have Never Been Modern*. Citing and analyzing historical examples, Latour questions the dualistic distinctions between nature and society, human and object, and ultimately, the distinction between the modern and the pre-modern. He calls for considering the (proliferation of) hybrids, which allows examinations of science and modernity to consider the intrinsic and deeply permeating networks linking nature, cultural elements, institutions, economies, and the state, and the continuous coexistence—instead of the revolutionary rupture—between the modern and the pre-modern.[31] To presume the autonomy of scientific disciplines from social issues by separating nature and society, or science and politics, is thus problematic not only in the case of Mao-era China.[32] In addition, the presumed neutrality and factuality of scientific disciplines seem to grant modern sciences a claim to truthfulness, which, as Latour argues in his *Politics of Nature* (2004), belongs to the domain of "the question of the common good" (values). The surreptitious confusion of values and facts in modern sciences not only conceals the stages of fabricating facts and the indispensable role of "theory" or "paradigm" that shape data, but also has the political consequence of short-circuiting "proper forms of discussion" in public life.[33]

The disunity of science highlighted in the introduction led to, among others, reflections of science in postcolonial studies. With the critique of a universalized modernity modeled on the historically contingent experiences of the West,[34] historians of science have tried to develop a discourse of science that takes into consideration notions of race, class, gender, and imperialism. Sandra Harding, for example, criticizes "the narratives of exceptionalism and triumphalism" of Western sciences, which idealize their capability of grasping reality in their own terms and tells a story of achievements with no significant downsides.[35]

The field of the history of science in China has experienced similar epistemological reflections and methodological modifications in the past decades. In his trailblazing series *Science and Civilisation in China*, Joseph Needham (1900–95) raised the famous question why science and technology in China stopped developing in the sixteenth century. In his observation, an important factor—among others—was that the socio-economic structure of imperial China prohibited the emergence of a merchant class that could have financed and benefited from the development of science and technology.[36] Nevertheless, he saw China was able to contribute to the genesis of modern, universal science in the seventeenth century when European and Chinese science merged. He described this process metaphorically as follows: "the older streams of science in the different civilizations like rivers flowing into the ocean of modern science."[37] The question we want to address in this book, to use Needham's metaphor, is not how these streams contributed to the unified ocean (Needham's assumption of the universality and rationality of science has received critiques, among others, for naturalizing power relations among civilizations[38]), but rather how streams merge, bringing with them different forms of sediment. Refusing to see science as "universal, objective, and value-free," Nathan Sivin reveals the ineffectiveness of understanding the history of science in China by rigidly applying European frames of reference such as the Scientific Revolution. Such a method, Sivin says (anticipating Harding's critique in 2008), not only fails to historicize the category of "science," but also holds "disastrous assumptions" that "turn the history of world science into a saga of Europe's success" and hence leads to the failure to comprehend "on their own terms technical inquiries of non-Western cultures."[39] Benjamin Elman's study *On Their Own Terms: Science in China 1550–1900*

(2005) and textbook *A Cultural History of Modern Science in China* (2006) foreground the importance of finding a valid "conceptual grid" to "explore Chinese interests in natural studies as they articulated and practiced them on their own terms rather than speculate about why they did not accomplish what the Europeans did"[40] and of writing a nuanced account of the "native vicissitudes" of science in China.[41]

These theoretical developments offer the possibility of viewing "science" as a historical form of knowledge which is not necessarily attached to modernity and/or modernization. And in doing so, we have a good chance of examining the truth-claim of science under concrete historical and social circumstances and avoiding the pitfalls of turning scientific knowledge production and dissemination into either a colonial project or a question of possible/necessary Westernization (or a total rejection of it). The growing number of recent publications dealing with the history of science and technology in China shows that science is no longer an abstract category tied to the assumption of modernity theory, but has become more flexible and encompasses a larger variety of practices and forms of knowledge.[42] The task of this book is to discuss the various contributions of the science discourse in the Mao era to the pluralization of knowledge in contemporary China.

DEFENDING "CORRECT SCIENCE" AGAINST SUPERSTITION IN THE MAO ERA

The party state in the Mao era has often been viewed as a monolithic existence, where the CCP's intellectual and ideological control intervened with the production and dissemination of scientific knowledge. Given the political persecution of scientists in twentieth-century China, a strong argument can be made to demand that science should ideally be independent from the political. Yet, a more important task of historical research is to reveal the many complexities in the trajectory of how "correct science" was defined, when it was presented in the public/official discourse as the effective tool to 1) improve the living standards of the people, 2) flatten the social hierarchy, 3) achieve China's political and economic independence, and 4) promote local experience and knowledge. We approach the definitions of "correct science" in the Mao era by considering some general questions: What constitutes "science?" Who counts as scientific authority? What is the correct approach to obtaining scientific knowledge? How should "science" be practiced, transmitted, and by whom?

Two editorials written by Mao Zedong in the Party newspaper, *Liberation Daily* (*Jiefang ribao* 解放日报), published respectively in 1941 and 1942, set up the keynote of "science" in the PRC. "Promoting Natural Sciences"

(*Tichang ziran kexue* 提倡自然科学) defined natural sciences as "the weapon to explain and conquer nature," and "Rectify the Party's Style of Work" (*Zhengdun dang de zuofeng* 整顿党的作风) stressed their function as "the knowledge about the struggle of production."[43] Therefore, natural sciences were seen as key to understanding nature and making it serve human needs and economic production.

These essays were written in the wake of the Natural Science Popularization Movement (*Ziran kexue dazhonghua yundong* 自然科学大众化运动) that had been launched in 1940 in the Shan-Gan-Ning Border Base during the Second Sino-Japanese War (1937–45). In the first editorial, Mao advocated the popularization (*tongsuhua* 通俗化) of science and scientific research, emphasizing the use of natural sciences to improve technologies of economic production, and calling for using dialectical materialism to replace the "metaphysical world view" of the bourgeois class in order to make natural sciences "thoroughly scientific" (*chedi de kexue* 彻底的科学). In the second editorial, Mao reiterates his epistemology of practice that combines both perceptual and rational stages of knowledge. While his targets appear to be both the dogmatists of Marxist theory and those overstressing one's experience, the emphasis of his disapproval lay in the former who only possessed bookish knowledge. Positing that Marxism-Leninism is "science leading proletarian revolutions to victory," Mao highlighted the importance of employing the viewpoint and methods of Marxism-Leninism to study concrete circumstances of Chinese revolution and to create theories adapting to its reality.

The ideas of "science" formulated in these two editorials hold several interrelated implications for the culture of knowledge in the first three decades of the PRC. In terms of the definition of "science," Marxism-Leninism—especially dialectical materialism—was viewed as both scientific knowledge and scientific method, and science was seen as applied knowledge that should produce tangible results in economic production. Therefore technology, "a system of knowledge and equipment that allows more or less efficient production of material goods and control over the environment,"[44] had become an integral part of "science." The privileging of technology as useful knowledge over theoretical research foreshadows the Communists' promotion of peasants and workers as knowledge producers, because technology had often been referred to as the knowledge of crafts(wo)men with low social status in the Chinese literati tradition. Such definition of "science" consequently put into question the authority of professionally trained experts, whose theoretical knowledge—and possibly also their "metaphysical world view"—appeared insignificant if not suspicious.

These ideas of science were spelled out and written into the 1949 Common Program of Chinese People's Political Consultative Conference (*Zhongguo renmin zhengzhi xieshang huiyi gongtong gangling* 中国人民政治协商会议共同纲领), which characterized culture and education as national, scientific, and of the masses (*dazhong de* 大众的). As part of culture and education, "science" was specified as natural sciences in service to industry, agriculture, and national defense as well as the "scientific historical view" to study and explain history, economics, politics, culture, and international affairs. "The love of science" (*ai kexue* 爱科学) was promoted, together with the love of motherland, the love of the people, the love of (manual) labor, and the care of public property, as a civic virtue (*gongde* 公德).[45]

A profound transformation in (scientific) knowledge, that is, what should be considered as valid knowledge and how knowledge should be produced and passed on, took place through the stormy criticism of the film *The Life of Wu Xun* (*Wu Xun zhuan* 武训传, 1950), a film on an education reformer in the Qing dynasty, who sponsored free schools with the money he accumulated through begging. The critique was directed against the film's praise of Wu Xun's reformist efforts which would question the historical necessity of revolution. Zhou Yang's article published in the *People's Daily* in August 1951 diagnosed the film as showing an "anti-people and anti-scientific view of history." The scientific approach to history and art, Zhou posited, should be socialist realism which saw the people and their labor as the driving force of historical progress.[46] Here Marxism-Leninism stood on its own as science dealing with both nature and society. Taking advantage of the cultural authority of science (dealing only with facts) established in the Republican time, the CCP was able to use the truth-claim of its "scientific" political ideology to champion its own management of knowledge.

In the early 1950s the ideal of "the People's Science" (*renmin kexue* 人民科学), which we will elaborate on in the next chapter, followed the example of the Soviet Union to seek an accelerated modernization of the nation. Upholding historical materialism, it took "scientific and industrial revolution as a natural outcome of human progress"[47] and expected the "scientific" political system of socialism would be able to manage and plan "science" more efficiently and progressively than in the capitalist world. Hundreds of model laborers (*laodong mofan* 劳动模范) were selected and publicized to exemplify the Socialist new (wo)man, who embodied a new human excellence combining scientific/technological competence, political loyalty, and moral selflessness, that is, a new generation of knowledge producers and disseminators that did not pursue science for the science's sake.[48]

Sharing the optimism of socialist transformation Zhou Enlai's report "On the Problem of Intellectuals" in January 1956 adjusted the official attitude toward intellectuals and theoretical knowledge. Aware of problems in the fast economic development in the on-going first Five-Year Plan (1953–57), it assigned intellectuals into the working class and called for more attention to fundamental (as opposed to applied) science. "Without theoretical research in science as the foundation," stated the report, "there will be no essential progress and innovations in technology." The acceleration of socialist industrialization, the report said, needed not just technological improvement for economic production, but also a full development of science and use of scientific knowledge. The report called for the campaign of "Marching towards Science" (*xiang kexue jinjun* 向科学进军) to cultivate China's own talents—by increasing the number of researchers and offering more continuing education opportunities for those holding positions.[49] It gestured toward the acknowledgment of professional—and theoretical—knowledge in the modernization process of the PRC and put professionals in a precarious position of being useful yet not authoritative. Subsequently a Committee of Scientific Development Planning (*Kexue guihua weiyuanhui* 科学规划委员会) was set up in March 1956 to draw the twelve-year "Prospective Plan for the Development of Science and Technology between 1956 and 1967" (1956–1967 *nian kexue jishu fazhan guihua gangyao* 1956–1967 年科学技术发展规划纲要).[50]

China's plan to cultivate its own talents to pursue a self-reliant modernization was catalyzed by Nikita Khrushchev's (1894–1971) De-Stalinization that started in February 1956. Deemed by the CCP as a revisionist policy, the leadership in Beijing embarked on a search for a different socialist modernity. Mao Zedong's April 1956 speech "On the Ten Major Relationships" (*Lun shida guanxi* 论十大关系) emphasized that one should not copy everything blindly and transplant mechanically from the Soviet Union. When the Hundred Flowers Campaign (*Baihua qifang* 百花齐放) started in May 1956 it claimed as its core principle the freedom to debate: "Letting a hundred flowers blossom and letting a hundred schools of thought contend" (*Baijia zhengming* 百家争鸣). The head of the Propaganda Department of the Central Committee of the Party Lu Dingyi 陆定一 (1906–96) assured his audience—natural scientists, social scientists, writers, and doctors—in a speech that the relationship between science (and art) and politics should not be oversimplified as a straightforward one.[51]

Before Maoist orthodoxy was re-established in the Anti-Rightist Movement (1957–59), however, the temporary reduction of political intervention in science made it possible to criticize Lysenkoism, part of the most controversial knowledge imported from the Soviet Union. Lysenkoism, perhaps the most studied example of Soviet science,[52] rejected Mendelian genetics in favor of hybridization theories similar to the Lamarckian theory of the inheritance of acquired characteristics. Trofim Lysenko (1898–1976) rose to power with the support of Stalin in 1948 and became a Soviet cultural hero and then a transnational icon of socialist science. He and his work were viewed as exemplifying Soviet proletarian biology rooted in the practical experience of the Russian—in contrast to Western—tradition. Lysenkoism was considered to be compatible with Marxist-Leninist thoughts, applicable to increasing agricultural productivity, and therefore showing "the ability of Soviet science to revolutionize man's relationship to the natural world."[53] When Lysenkosim arrived in China in 1952, other genetic theories were banned in the People's Republic.

The fact that Lysenko had already come under attack by Soviet biologists in the same year was unknown to Chinese scientists. While the centenary of Iwan W. Michurin (1855–1935), the plant breeder lauded by Lysenko as a scientific hero with practical experience, was still celebrated in China in 1955, the situation changed in 1956 when Hans Stubbe (1902–89), the president of the East German Academy of Agricultural Science, lectured at Beijing Agricultural University and told his audience that there was no scientific foundation for Lysenkoism.[54] In April 1956, the Soviet academician Nikolai Vasilyevich Tsytsin (1898–1980),[55] a specialist in biology and agriculture, told his Chinese colleagues that Lysenko had been dismissed from his duties. On the two-week Genetics Symposium held in Qingdao in August of that year Tan Jiazhen 谈家桢 (1909–2008), an academician of the Chinese Academy of Science who obtained his Ph.D. in 1937 at the California Institute of Technology while working with Thomas Morgan on the fruit fly (drosophila), boldly stated that Lysenkoism was not scientific and that socialist genetics had suffered regression due to the lack of the access to the latest developments in DNA and molecular biology research.[56] The historical significance of the Qingdao Symposium lies in the fact that the Soviet Union ceased to be the exclusive knowledge provider for China's modernization, and some defendants of Morganism even saw the Soviet knowledge as "colonial science."[57] The vicissitude of Lysenkoism refracted how the young PRC floundered to build its own socialist science by defining "science" and scientific methods. At that time, new sources of scientific knowledge came into focus, be they science and technology from other states in the Eastern bloc than the Soviet Union,[58] or native sources of knowledge as we are going to show in the course of this book.

With the begin of the Anti-Rightist Campaign, the authority of science became blurred again. When "Mass science" (*qunzhong kexue* 群众科学) emerged as a new approach, a volume titled *A Collection of [the Results of] Mass Scientific Research* (*Qunzhong kexue yanjiu wenji* 群众科学研究文集, 1958), for example, collected editorials of local newspapers and short reports on scientists and experts engaging in practical research and learning from the masses.[59] In June 1958, *People's Daily* published two editorials announcing, respectively, a "revolution in technology (*jishu geming* 技术革命)" and a "revolution in culture (*wenhua geming* 文化革命)." The "revolution in technology" aimed to realize the industrialization of the country by mobilizing a mass movement of technical improvement and technological innovation. More specifically, the goals were to realize mechanization and electrification, following the general principle of "more, faster, better, and more economical" (*duo kuai hao sheng* 多快好省).[60] To make the "revolution in technology" happen, a "revolution in culture" must be carried out to allow the masses to learn reading and science.[61]

These editorials established the masses as the agent of knowledge production. Closely following these calls for revolutions came the official vision of "walking on two legs" (*liangtiaotui zoulu* 两条腿走路) as the proper method of knowledge production. In early 1958, this term referred to the school reform of founding private agricultural schools, but with a *People's Daily* editorial published on June 18, 1958, it became a catchphrase describing the practice-oriented mass line of creating scientific and technological knowledge that emphasized using local human and material resources to increase economic productivity. "As a matter of fact, science comes from the practice of class struggle and [economic] production," claimed the editorial, "there is nothing mysterious about it; the brightest and most talented people in the world are more often than not those with most practical experiences (*shijian jingyan* 实践经验)." It further attributed "important inventions in the world history" to "those mostly with lower social class, less scholarship, and poorer conditions," because they possessed "rich practical experiences." Walking on two legs, therefore, meant to depend on "both the central government and the local government, both on experts and, even more, on the masses."[62] When publishing his article on mass science in the Party organ *Red Flag* (*Hongqi* 红旗) in 1959, the peak time of the Great Leap Forward campaign, Zeng Xisheng 曾希圣 (1904–68), then the Party secretary of Anhui Province where the worst famine took place, used "walking on two legs" to refer to both foreign/modern (*yang*) and indigenous (*tu*) methods to realize further economic development.[63]

If "walking on two legs" were a proper approach to knowledge to empower the masses, then it should have led to the reduction and decentralization of the state's control of science and technology. This, however, did not happen. On the contrary, the central planning was held as the key tenet of socialist science to avoid waste of resources. Nie Rongzhen 聂荣臻 (1899–1992) stated in 1958 that science could certainly be planned and that socialist states should not allow freedom in scientific research as capitalist societies do.[64] In other words, the production of knowledge by the masses was a political project that needs strict coordination, guidance, and supervision of the ideological avant-garde; and consequently, the goal of empowering workers and farmers ended up bringing in more state intervention.

With regard to knowledge production, however, the practice-based and production-oriented approach to knowledge allowed for a greater degree of heterogeneity, as we will show in the following chapters. Liu Xianzhou 刘仙洲 (1890–1975), the vice president of Tsinghua University then, opened his 1959 article titled "Science and Economic Production" with the following statements: "The term science originally meant 'knowledge' in European languages. And knowledge comes from the practice of production activities (科学这一个名词，在欧洲文字中的原义是"知识"。而知识的基本来源是人们生产活动的实践)." By going back to the Chinese history of agriculture, Liu argued for the necessary connections between science, technological inventions, and the increase of productivity by attributing the development of science, above all, to the needs of economic production. The theoretical part of science, according to Liu, must found itself on the results of practice of production or experimentation, and in turn, be verified by them. After this deft application of Mao Zedong's epistemology of practice, Liu differentiated socialist and capitalist sciences in the last part of his article. Socialist science had an interactive—hence presumably democratic—relations to the non-specialists: it opened itself for the use of the people while accepting their contribution; capitalist science, on the other hand, was obsessed with the pursuit of profit and therefore limited its own use and ultimately also it development.[65] Liu seemed to attempt to valorize professional science in the Great Leap years by stressing its empirical process of knowledge production and interactions with the masses.

The emphasis on practice and the critique of the authority of the expert contributed in the Great Leap Forward era to the fight against "superstition." It was not a fight trying to dispel wrong knowledge, but rather one to dispel the myth of the expert and his power—fundamentally it dealt with the issue of class struggle. In her analysis of the concerted efforts of political and scientific elites to abolish religious forms of knowledge as superstitious in the early PRC time, Schmalzer insightfully comments on the intriguing consequences that the intellectuals had to bear later: "In attacking religious forms of knowledge, scientists helped strengthened the authoritarian character of the state ideology," and the "implicit elitism of discourse on superstition lent weight to the perception that intellectuals—including scientists—represented a class divorced from the 'broad masses of the people'" that would fuel "the kind of bottom-up, mass science correctives that characterized the more radical periods of the People's Republic."[66]

Indeed, in the radical period of the Great Leap Forward, the political ideology of class struggle underlay Mao Zedong's epistemology of practice to destroy pseudoscience (*wei kexue* 伪科学), as shown in the title of a 1958 article: "Practice smashed the pseudo-science of the bourgeois class" (*Shijian fensuile zichanjieji de weikexue* 实践粉碎了资产阶级的伪科学). This article argued that the false belief in expert authority (*jishu quanwei* 技术权威) installed by the bourgeois class hindered economic production, and the bourgeois' attempt to maintain this authority was to preserve their privileged class status.[67] An article on the dichotomy of science and superstition in *Red Flag* further opposed one mode of scientific thinking to the other—the superstitious one.[68] According to this dichotomy, there was only one mode of scientific thinking that is correct, and others were plainly wrong. Following this dichotomous line, superstition would enter a pure political arena in 1961, when the CCP's real and imagined enemies—imperialism and reactionaries— were metaphorically referred to as "ghosts," a term evoking the popular religious God-demon dichotomy, and hence were conveniently categorized into unscientific—and harmful—"superstition."[69] Banishing "false/wrong" knowledge was, however, not an easy task. Contending the effectiveness of anti-superstitious campaigns in the Mao era, Denise Ho argues that they often relied on "a rhetoric of science *qua* science than on scientific inquiry." Science simply instilled in the students as "a set of incontrovertible facts and fixed principles" that excluded other ways of thinking and methods failed to convince the students, who only obeyed the teacher "with their mouths and not with their hearts."[70]

In view of these observations it seems wise not to overestimate the power of propaganda in the Mao era. The definitions of "correct science," and correspondingly, "superstition," as shown above, were not consistent and stable. The epistemologies of knowledge, while largely sticking to Mao's tenet of "practice," were interpreted in various ways at different historical moments for different purposes. These inconsistencies in the public discourse of science, while certainly indicating chaos in many aspects of the PRC, also opened up the possibilities of pluralizing knowledge in actual scientific and technological practices. In the next chapter, we will examine social processes of creating, legitimizing, and disseminating scientific knowledge in the first decades of the PRC as well as the integrated and interactive role that science played in legitimizing the Party's political and social ideals. We explore how the "people's science" came into being in the larger historical context of the Cold War and what cognitive strategies were employed to convey the idea of correct science vis-à-vis superstition.

NOTES

1. According to Chen Hongjie, the title of this magazine was inspired by the American popular science magazine *Scientific American* (founded in 1845). See Chen Hongjie 陈洪杰, *Zhongguo jindai kepu jiaoyu: shetuan, changguan he jishu* 中国近代科普教育：社团、场馆和技术 (Modern science education in China: societies, spaces, and technologies) (MA thesis, East China Normal University, 2006), 21.

2. Zhang Qiyun 张其昀, "Zhongguo kexuehua yundong xiehui faqi zhiqu shu 中国科学化运动协会发起旨趣书 (The objectives of the Association for the China Scientization Movement)," *Kexue de Zhongguo* 科学的中国 1, no. 1 (1933): 1–3.

3. Nathan Sivin, *Chinese Alchemy: Preliminary Studies* (Cambridge, MA: Harvard University Press, 1968); Nathan Sivin, *Traditional Medicine in Contemporary China: A Partial Translation of Revised Outline of Chinese Medicine (1972) with an Introductory Study on Change in Present-day and Early Medicine* (Ann Arbor: Center for Chinese Studies, University of Michigan, 1987); Pingyi Chu, "Narrating a History for China's Medical Past: Christianity, Natural Philosophy and History in Wang Honghan's *Gujin yishi* 古今醫史 (History of Medicine Past and Present)," *EASTM*, no. 28 (2008): 14–35; Rebecca Nedostup, *Superstitious Regimes: Religion and the Politics of Chinese Modernity* (Cambridge: Harvard University Press, 2009); Sean Hsiang-lin Lei, *Neither Donkey nor Horse: Medicine in the Struggle over China's Modernity* (Chicago: University of Chicago Press, 2014).

4. Stephan Feuchtwang, "The Problem of 'Superstition' in the People's Republic of China," in *Religion and Political Power*, eds. Gustavo Benavides and Martin W. Daly (Albany: State University of New York Press, 1989), 43–68; Ole Bruun, *Fengshui in China: Geomantic Divination between State Orthodoxy and Popular Religion* (Honolulu: University of Hawai'i Press, 2003); Steven A. Smith, "Local Cadres Confront the Supernatural: The Politics of Holy Water (*Shenshui*) in the PRC, 1949–1966," *China Quarterly*, no. 188 (2006): 999–1022; Steven A. Smith, "Talking Toads and Chinless Ghosts: The Politics of 'Superstitious' Rumors in the People's Republic of China, 1961–1965," *The American Historical Review* 111, no. 2 (2006): 405–27; David Palmer, *Qigong Fever: Body, Science, and Utopia in China* (New York: Columbia University Press, 2007).

5. Taken from Wang Zuoyue, "Saving China through Science: The Science Society of China, Scientific Nationalism, and Civil Society in Republican China," *Osiris*, 2nd Series, 17, Science and Civil Society (2002): 291–322, here p. 309. The original quote can be found in Hu Shi 胡适, "Kexue yu renshengguan xu 科学与人生观序 (Preface to Science and the View of Life)," in *Kexue yu renshengguan* (Science and the View of Life), ed. Zhang Junmai (Jinan: Shandong People's Press, 1997), 12.

6. Arthur Smith, *Chinese Characteristics* (New York: Fleming H. Revell, 1894), 48–49, 50.

7. This point was also made by Daston and Galison who have shown that objectivity is not an ontological given, but has a history of itself, see Lorraine Daston and Peter Galison, *Objectivity* (New York: Zone Books, 2010).

8. Chaosheng 潮生, "Shishi qiushi 实事求是 (Seeking Truth from Facts)," *Tongwen yuekan* 通问月刊, no. 4 (1931): 1–2.

9. Mao's advocacy of practice-oriented, fact-based epistemology of knowledge may partly explain why the fierce attacks on Hu Shi and his teacher, the philosopher of American pragmatism John Dewey (1859–1952), by communist cadres such as Guo Moruo 郭沫若 (1892–1978), Lu Dingyi and Zhou Yang 周扬 (1907–89) during the 1950s focused mainly on Hu's rejection of historical materialism and on Dewey as class enemy. Hu Shi's view of science, especially that of the role of "experience" (*jingyan* 经验) in understanding the world in the tradition of Dewey's philosophy, was on the other hand forcefully defended and promoted. On a brief yet comprehensive summary on the Communist attack on Hu Shi see Jerome Grieder, *Hu Shih and the Chinese Renaissance: Liberalism in the Chinese Revolution, 1917–1937* (Cambridge, MA: Harvard University Press, 1970), 358–68. With the May Fourth era also coinciding his own formative years, Mao Zedong's idea of practice might possibly be influenced by Dewey's pragmatism.

10. Detailed studies on the history of how the symbol of the Sick Man of East Asia was used in twentieth-century political discourses offer Yang Ruisong 楊瑞松, *Bingfu, huanghuo yu shuishi: "Xifang" shiye de Zhongguo xingxiang yu jindai Zhongguo guozu lunshu xiangxiang* 病夫、黃禍與睡獅:"西方"視野的中國形象與近代中國國族論述想像 (Sick man, yellow peril and sleeping lion: the image of China in the eyes of the West and the narrative imagination of the modern Chinese nation) (Taibei: Zhengda chubanshe, 2010); and Yang Nianqun 杨念群, *Zaizao "bingren": Zhong-Xi yi chongtuxia de kongjian zhengzhi* 再造"病人":中西医冲突下的空间政治 (Remaking "Patients": Spacial Politics in the Conflicts between Eastern and Western Medicine) (Beijing: Zhongguo Renmin daxue chubanshe, 2013).

11. See here Theodore M. Porter, *Trust in Numbers: The Pursuit of Objectivity in Science and Public Life* (Princeton: Princeton University Press, 1995). For the case of Republican China see Daniel W. Y. Kwok, *Scientism in Chinese Thought 1900–1950* (London: Yale University Press, 1965).

12. This section has profited from the research of Lei, *Neither Donkey nor Horse*; Sivin, *Chinese Alchemy*; John Bower, William Hess, and Nathan Sivin, *Science and Medicine in Twentieth-Century China: Research and Education* (Ann Arbor: Center for Chinese Studies, University of Michigan, 1988); Volker Scheid, *Chinese Medicine in Contemporary China: Plurality and Synthesis* (Durham: Duke University Press, 2002); Kim Taylor, *Chinese Medicine in Early Communist China, 1945–1963: A Medicine of Revolution* (London: Routledge, 2005).

13. It would be misleading to translate *Zhongyi* as traditional Chinese medicine (or TCM), because TCM is a name created for foreign consumption, as shown by Taylor, *Chinese Medicine in Early Communist China*. On TCM as an invented tradition see Elisabeth Hsu, "The History of Chinese Medicine in the People's Republic of China and Its Globalization," *East Asian Science, Technology and Society* 2, no. 4 (2008): 465–84. According to her, "TCM, in spite of being called 'traditional' (*chuantong*), is generally referred to as the 'modernised' (*xiandaihuade*), 'scientific' (*kexuehuade*), 'systematic' (*xitonghuade*), and 'standardised' (*guifanhuade*) Chinese medicine." See Elisabeth Hsu, *The Transmission of Chinese Medicine* (Cambridge: Cambridge University Press, 1999), 7.

14. A very instructive overview on this issue is provided by Xu Xiaoqun, "'National Essence' vs 'Science': Chinese Native Physicians' Fight for Legitimacy, 1912–37," *Modern Asian Studies* 31, no. 4 (1997): 847–77. The most well-known case of such personal transformation was Lu Xun who in his younger years—as described in the preface to his work *Call to Arms* where he recalls that after going to pawn shops to raise money to purchase medical prescriptions for his father that didn't take effect—started to despise "traditional medicine" and other forms of "superstitions." This was one of the reasons why he studied modern medicine in Japan. See his autobiographical account in the preface to *Call to Arms* (*Nahan* 呐喊), in Lu Xun (Lu Hsün), *Complete Stories*, trans. Yang Xianyi (Bloomington: Indiana University Press, 1981). The rejection of CM by Sun Yat-sen is detailed by Lei, *Neither Donkey nor Horse*.

15. On the attack on superstitious practices in the medical field see also Nedostup, *Superstitious Regimes*, 191–226.

16. For this argument see Steven Smith, "Introduction: The Religion of Fools? Superstition: Past and Present," *Past and Present* 199 (2008): 7–55, here p. 9.

17. Yu Yunxiu, "How to Popularize Scientific Medicine in China," *Shenbao Medical Weekly*, no. 111 (1935). Translation taken from Lei, *Neither Donkey nor Horse*, 45. On the role of Yu Yunxiu see further Taylor, *Chinese Medicine in Early Communist China*. According to Lei, Yu Yunxiu also preferred the local Chinese medicine of the pre-Song era (960–1279) over the academic, theorized medicine of literati-scholars of later dynasties, see Sean Hsiang-lin Lei, "How Did the Chinese Medicine Become Experiential? The Political Epistemology of Jingyan," *positions* 10, no. 2 (2002): 333–64. Nathan Sivin and William Cooper have refuted the idea of an invented empirical tradition in Chinese medicine: even before the institutionalization of Chinese medicine in the Song dynasty theory played a role. See William Cooper and Nathan Sivin, "Man As a Medicine: Pharmacological and Ritual Aspects of Traditional Therapy Using Drugs Derived from the Human Body," in *Chinese Science: Exploration of an Ancient Tradition*, ed. Shigeru Nakayama and Nathan Sivin (Cambridge: MIT Press, 1973), 203–72, esp. 206.

18. Ruth Rogaski, *Hygienic Modernity: Meanings of Health and Disease in Treaty-Port China* (Berkeley: University of California Press, 2004).

19. Lei, *Neither Donkey nor Horse*, 102.

20. Lei, *Neither Donkey nor Horse*, 4.

21. Lei, *Neither Donkey nor Horse*. In a similar way V. Scheid points out that that neither system is a systematic agglomeration of non-contradictory knowledges and practices. Arguing for a medical pluralism he points to the fact that the search for health is a "dynamic, discontinuous, and fragmented process involving complex negotiations of social identity and morality" (Scheid, *Chinese Medicine in Contemporary China,* 11).

22. Lü Guangming 吕光明, "Sulian fangyi renyuan di Zhang-shi–Zhong-Su ge fangyidui fenfu Cha Meng yiqu Cha-bei jumin mixin sixiang jidai kefu 苏联防疫人员抵张市—中苏各防疫队分赴察蒙疫区察北居民迷信思想急待克服 (Soviet plague prevention workers arrived in the city of Zhangjiakou—Chinese and Soviet teams went to infected areas in Chahar and Mongolia, residents in north Chahar need to overcome their superstitious thoughts)," *Renmin Ribao*, November 7, 1949.

23. Lü Guangming 吕光明, "Zhang-shi baiyu yiwu renyuan fangzhi shuyi jin shiwan ren shou zhushe—xuexiao jiaoyuan jinxing pochu mixin xuanchuan 张市百余医务人员防治鼠疫近十万人受注射—学校教员进行破除迷信宣传 (More than one hundred medical workers in the city of Zhangjiakou prevent and treat bubonic plague, nearly 100,000 people received injections—school teachers join the propaganda against superstition)," *Renmin Ribao*, November 11, 1949.

24. Smith, "Local Cadres Confront the Supernatural," quotes numerous cases to argue these points. For the fight against religious superstition as economic offense in the 1980s see also Feuchtwang, "The Problem of 'Superstition' in the People's Republic of China."

25. Tang Shenglan, Gerald Bloom, Xushen Feng, Henry Lucas, Xingyuan Gu, and Malcom Segall, with Gail Singleton and Polly Payne, "Financing Health Services in China: Adapting to Economic Reform" (Brighton: Institute for Development Studies, 1994), here quoted after Scheid, *Chinese Medicine in Contemporary China*, 67–68.

26. The situation changed only in 1953 after the intervention of Mao Zedong who acknowledged the lack of sufficient health care for all levels of society and called for a reduced dependency on foreign contributions. For an overview on the institutionalization of human medicine in the Mao era, see Scheid, *Chinese Medicine in Contemporary China*, 67–81.

27. Zhongyang weishengbu weisheng xuanchuanchu 中央卫生部卫生宣传处, "Yufang xiaji chuanranbing 预防夏季传染病 (Preventing Contagious Diseases in the Summer Season)," *Kexue puji gongzuo*, no. 16 (June 1951), 179–82.

28. Recent scholarship has highlighted epistemological complexities in communicating knowledge in contexts where "seeing is believing" failed to work, for example, when knowledge about subtle, microscopically small and highly fluid objects—be it snowflakes, bacteria, or protons and neutrons—was communicated to non-expert audiences that lacked a similar cognitive base. See Gross, *Farewell to the God of Plague;* and Matten, "Coping with Invisible Threats."

29. Rogaski, *Hygienic Modernity*.

30. Fan, "East Asian STS: Fox or Hedgehog?"; Anderson, "Postcolonial Specters of STS." On self-assertion as discourse in East Asia see Iwo Amelung, Matthias Koch, Joachim Kurtz, Eun-Jeung Lee, and Sven Saaler, eds., *Selbstbehauptungsdiskurse in Asien: China—Japan—Korea* (München: Iudicium, 2003).

31. Bruno Latour, *We Have Never Been Modern*, trans. Catherine Porter (Cambridge, MA: Harvard University Press, 1993).

32. Even only for the twentieth century the instances where politics intervened in science and knowledge production are uncountable, for example, to the Lysenko affair (Shores A. Medvedjev, *The Rise and Fall of T. D. Lysenko* [New York: Columbia University Press, 1971]); Schneider, *Biology and Revolution in Twentieth-Century China*); the critique of Albert Einstein's theory of relativity (Hu, *China and Albert Einstein*), or the persecution of mathematicians (Huang, "Youpai wenxuezhong de zirankexuejia").

33. Bruno Latour, *Politics of Nature: How to Bring the Sciences into Democracy*, trans. Catherine Porter (Cambridge, MA: Harvard University Press, 2004), 93.

34. See, for example, Dipesh Chakrabarty, *Provincializing Europe: Postcolonial Thought and Historical Difference* (Princeton: Princeton University Press, 2008).

35. Harding, *Sciences from Below*, 3–4.

36. Joseph Needham, "Science and Society in East and West," *Science and Society* 28, no. 4 (1964): 385–408. Economic historians argued that—given a society with labor surplus—economic growth could only be achieved by demographic growth, resulting in the high-level equilibrium trap. See here the works by Mark Elvin and Christopher Mills Isett, such as Mark Elvin, "The High-Level Equilibrium Trap: The Causes of the Decline of Invention in the Traditional Chinese Textile Industries," in *Economic Organization in Chinese Society*, ed. W. E. Willmott (Stanford: Stanford University Press, 1972), 137–72; Kenneth Pomeranz, *The Great Divergence: China, Europe and the Making of the Modern World Economy* (Princeton, NJ: Princeton University Press, 2001); and Christopher M. Isett, *State, Peasant, and Merchant on the Manchurian Frontier, 1644–1862* (Stanford: Stanford University Press, 2007).

37. Joseph Needham, "The Roles of Europe and China in the Evolution of Oecumenical Science," *Journal of Asian History* 1, no. 1 (1967): 4. Characteristic for the universalizing notion of science are questions such as "Why did China not develop modern science (as did Europe)?," cf. here Justin Yifu Lin, "The Needham Puzzle: Why the Industrial Revolution Did Not Originate in China," *Economic Development and Cultural Change* 43, no. 2 (1995): 269–92.

38. See Chu Pingyi, "Narrating a History for China's Medical Past." Andre Gunder Frank has argued that Needham, in his efforts to provide a full picture of China's technological history, never fully abandoned his preconceived notions of European exceptionalism. See Andre Gunder Frank, *ReORIENT. Global Economy in the Asian Age* (Berkeley and Los Angeles, California: University of California Press, 1998).

39. Nathan Sivin, "Why the Scientific Revolution Did Not Take Place in China—or Didn't It?," last accessed January 18, 2021, http://ccat.sas.upenn.edu/~nsivin/writ.html, 6–14 (revised version of an essay published 1982 in *Chinese Science*, no. 5 [2005]: 45–66).

40. Benjamin A. Elman, *On Their Own Terms: Science in China, 1550–1900* (Cambridge: Harvard University Press, 2005), xxvi.

41. Benjamin A. Elman, *A Cultural History of Modern Science in China* (Cambridge: Harvard University Press, 2006), 13.

42. See for instance Schmalzer, *Red Revolution, Green Revolution;* John Law and Wen-yuan Lin, "Provincializing STS: Postcoloniality, Symmetry, and Method," *East Asian Science, Technology and Society*, no. 11 (2017): 211–27.

43. Mao Zedong, "Tichang ziran kexue 提倡自然科学," *Jiefang Ribao*, June 12, 1941; Mao Zedong, "Zhengdun dang de zuofeng 整顿党的作风." The second essay was originally published as "Zhengdun xuefeng dangfeng wenfeng 整顿学风党风文风 (Rectify the Party's style of learning, working, and writing)," *Jiefang ribao*, April 27, 1942, then republished in Mao Zedong 毛泽东, "Zhengdun dang de zuofeng 整顿党的作风," *Mao Zedong zhuzuo xuandu* (*xiace*) 毛泽东著作选读 (下册) (Beijing: Renmin chubanshe, 1986), 487–505.

44. Francesca Bray, *Technology and Gender: Fabrics of Power in Late Imperial China* (Berkeley: University of California Press, 1997), 7.

45. "Zhongguo renmin zhengzhi xieshang huiyi gongtong gangling 中国人民政治协商会议共同纲领 (Common Program of Chinese People's Political Consultative Conference)," *Jiangxi zhengbao* 江西政报 (*Jiangxi Government Information*), no. 3 (1949): 19–20.

46. Zhou Yang 周扬, "Fan renmin, fan lishi de sixiang he fan xianshizhuyi de yishu: dianying *Wu Xun zhuan* pipan 反人民、反历史的思想和反现实主义的艺术：电影《武训传》批判 (Anti-people and anti-historical thought and anti-realistic art: criticizing the film *The Life of Wu Xun*)," *Renmin Ribao*, August 8, 1950.

47. Bray, *Technology and Gender*, 9.

48. In 1955, the physicist and vice president of Chinese Academy of Sciences (CAS) Wu Youxun 吴有训 (1897–1977) had warned against doing research in the sense of "la science pour la science" (*wei kexue er kexue* 为科学而科学), because doing science without considering practical application would be detrimental to socialist construction. See Wu Youxun 吴有训, *Zhongguo Kexueyuan wulixue shuxue huaxue bu baogao* (1955 *nian 6 yue 2 ri zai Zhongguo Kexueyuan xuebu chengli dahuishang de baogao*) 中国科学院物理学数学化学部报告 [1955年6月2日在中国科学院学部成立大会上的报告] (Report on the Department of Physics, Mathematics and Chemistry of the Chinese Academy of Sciences [Report at the founding meeting of the Chinese Academy of Sciences on June 2, 1955]), *Lun woguo de kexue gongzuo* 论我国的科学工作 (Discussing our country's scientific work) (Beijing: Renmin chubanshe, 1956), 61.

49. Zhou Enlai, "Guanyu zhishifenzi *wenti* de baogao 关于知识分子问题的报告 (On the Problem of Intellectuals)," *Renmin Ribao*, January 30, 1956.

50. "Zhongguo gongchandang dashiji. 1956 nian 中国共产党大事记. 1956 年," *News of the Communist of China* 中国共产党新闻, accessed November 15, 2018, http://cpc.people.com.cn/GB/64162/64164/4416035.html. On the plan see also Zuoyue Wang, "The Chinese Developmental State During the Cold War: The Making of the 1956 Twelve-year Science and Technology Plan," *History and Technology* 31, no. 3 (2015): 180–205.

51. Lu Dingyi, "Baihua qifang, baijia zhengming--yijiu wuliu nian wu yue ershiliu ri zai Huairentang de jianghua 百花齐放，百家争鸣--一九五六年五月二十六日在怀仁堂的讲话 (Let hundred flowers bloom and hundred schools of thoughts contend—a talk in Huairen Hall on May 26, 1956)," *Renmin Ribao*, June 13, 1956. English translation in Robert R. Bowie and John K. Fairbank, eds., *Communist China 1955–1959: Policy Documents with Analysis* (Cambridge: Harvard University Press, 1965), 151–63.

52. Ethan Pollock, *Stalin and the Soviet Science Wars* (Princeton: Princeton University Press, 2006), 230, fn. 1.

53. Pollock, *Stalin and the Soviet Science Wars*, chapter 3, quote 42. Regarding the reasons why Lysenkoism was so popular despite its many loop-holes, James Crow writes: "The strong appeal of Michurin-Lysenko genetics is not hard to understand. It bypasses the necessity to understand the complex behavior of chromosomes during meiosis. The intellectual rigor of chromosome mapping and the difficult techniques

for cytological study are not needed. There is no need to understand statistics. This enticed the ignorant and the lazy. Most importantly, it promised immediate results; there was no need to wait several generations for Mendelian selection to work." See James Crow, "Genetics in Postwar China," in *Science and Medicine in Twentieth-Century China*, eds., Bowers, Hess, and Sivin (Ann Arbor: Center for Chinese Studies, University of Michigan, 1988), quote 160.

54. After his own visit to the Lysenko Institute in Moscow, Stubbe had tried to reproduce the Soviet experiments at his institute in Gatersleben, yet failed to do so. See Johannes Siemens, "Lyssenkoismus in Deutschland (1945–1965)," *Biologie in unserer Zeit*, no. 27 (1997), 255–62; and Edda Käding, *Engagement und Verantwortung. Hans Stubbe, Genetiker und Züchtungsforscher. Eine Biographie* (Müncheberg, 1999).

55. From 1945 until his death Tsytsin had been the director of the Main Moscow Botanical Garden of Academy of Sciences (today named after him) and member of the Supreme Soviet of the Soviet Union. His main field of research was breeding of new crop varieties.

56. Tan Jiazhen 谈家桢, "Wo dui yichuanxue zhong yixie wenti de kanfa 我对遗传学中一些问题的看法 (My opinions on some problems in genetics)," *Renmin Ribao*, September 6, 1956. According to the *People's Daily* of October 7, 1956, both schools presented their recent discoveries one after the other before engaging in discussions. See Huang Qinghe 黄青禾 and Huang Shun'e 黄舜娥, "Yige chenggong de xueshu huiyi—ji yichuanxue zuotanhui 一个成功的学术会议——记遗传学座谈会," *Renmin Ribao*, October 7, 1956; and Li Peishan, Meng Qinzhe, Huang Qinghe, and Huang Shu-e, "The Qingdao Conference of 1956 on Genetics: The Historical Background and Fundamental Experiences," in *Chinese Studies in the History and Philosophy of Science and Technology*, eds. Robert S. Choen and Dainian Fan (Boston: Kluwer Academic Publishers, 1996), 41–54.

57. Schneider, *Biology and Revolution in Twentieth-Century China*, 176.

58. Austin Jersild, *The Sino-Soviet Alliance: An International History (New Cold War History)* (North Carolina: University of North Carolina Pres, 2014); Marc A. Matten, "Turning Away from the Big Brother: China's Search for Alternative Sources of Knowledge During the Sino-Soviet Split," *Comparativ* 29, no. 1 (2019): 64–90.

59. *Qunzhong kexue yanjiu wenji* 群众科学研究文集 (Beijing: Kexue puji chubanshe, 1958).

60. Editorial, "Xiang jishu geming jinjun 向技术革命进军," *Renmin Ribao*, June 3, 1958.

61. Editorial, "Wenhua geming kaishi le 文化革命开始了 (A revolution in culture has began!)," *Renmin Ribao*, June 9, 1958.

62. "Difang ye neng daban kexue shiye 地方也能大办科学事业 (Local government can also finance scientific development)," *Renmin Ribao*, June 18, 1958.

63. Zeng Xisheng 曾希圣, "Tan yangtu bingju 谈洋土并举 (On upholding both foreign and indigenous methods)," *Hongqi*, no. 6 (1959): 20–24.

64. Nie Rongzhen 聂荣臻, "Woguo kexue jishu gongzuo fazhan de daolu 我国科学技术工作发展的道路 (The way of developing science and technology in our

country)," *Hongqi*, no. 9 (1958): 4–15. Nie, military leader of the PLA, was made a Marshal in 1955 and was later responsible for the Chinese nuclear weapons program.

65. Liu Xianzhou, "Kexue yu shengchan 科学与生产," *Hongqi*, no. 9 (1959): 22–29.

66. Schmalzer, *The People's Peking Man*, 84–85. On the fight against superstition in the Mao era see also Gross, *Farewell to the God of Plague*; Denise Y. Ho, *Curating Revolution: Politics on Display in Mao's China* (Cambridge: Cambridge University Press, 2017).

67. Lin Liang 林亮, "Shijian fencuile zichanjieji de weikexue 实践粉碎了资产阶级的伪科学," *Chuangzao*, no. 2 (1958): 31–34.

68. Jiang Ke 江珂, "Kexue he mixin 科学和迷信 (Science and superstition)," *Hongqi*, no. 13 (1958): 21–24. On how to differentiate science and pseudoscience see Mario Bunge, "Demarcating Science from Pseudoscience," *Fundamenta scientiae*, no. 2 (1982): 369–88.

69. Institute of Literature of the Chinese Academy of Social Sciences, *Stories about Not Being Afraid of Ghosts*, trans. Yang Hsien-yi and Gladys Yang (Peking: Foreign Languages Press, 1961). For the politicization of demonology in popular religions, see Barend J. ter Haar, "China's Inner Demons: The Political Impact of the Demonological Paradigm," *China Information* XI, nos. 2/3 (1996–97): 54–88.

70. Ho, *Curating Revolution*, 125.

Chapter Two

Creating the People's Science
Science Dissemination as a Social Process

After the founding of the PRC in 1949, the CCP sought to develop the People's Science (*renmin kexue*), that is, a new—and presumably more effective and cost-saving—way of generating scientific knowledge and thinking. Appearing first in the early 1950s,[1] it was a key notion in Maoist science discourse. It highlighted science dissemination as a social process, in which the making of correct science was linked and interacted with the making of the people. People's science carried on the idea of "popular science" (*dazhong kexue* 大众科学) from the Republican era, which believed that the general reader's grasp of modern science and technology as well as evolutionary thinking would bring about a strong "scientific China."[2] In the late 1950s the People's Science would evolve into a more radical, populist version—"mass science" (*qunzhong kexue* 群众科学), which destabilized the authority of professional knowledge production by endorsing and promoting experience-based local knowledge and native technology of peasants and workers. What had remained constant in all these ideas and practices of science was the goal of creating qualified human resources equipped with proper scientific and technological literacy and skills to modernize China.

Given the constrained financial, natural, and human resources in the early years of the PRC, the People's Science was informed by a pragmatism that tried to activate the agency and energy of the large population. Its creation by the state, therefore, was carried out at multiple levels. It, first of all, transmitted basic knowledge of science and technology and the idea of their importance for a modern and self-reliant China; at the same time it tried to convince the populace of the universal truthfulness of science and the equally "scientific" political system of socialism with its social order and values; and last but not least, it created the vision of an empowered "people" of the new China by addressing ordinary peasants and workers as emancipated laborers full of creative potentials. It is, however, important to note that the creation of this ideal, in particular its search for alternative ways of non-expert knowledge making, gave rise to more confusions and risks than liberating potentials. It was, furthermore, in many cases imposed by means of political coercion. Yet its highly experimental nature has never been debated—or allowed to debate—in the public discourse of science in the Mao era.

This chapter delineates the social process of science dissemination in the 1950s and the 1960s, in which ideological agendas and social ideals were integrated. It first looks at the ideas, practices, and personnel in science dissemination across the 1949 divide and then examines its institutionalization in the early PRC, in which the expert was constantly repositioned in knowledge production and dissemination. The last parts analyze three non-textual forms of communicating scientific knowledge to the populace: exhibition, film, and magazine illustration. Instead of expecting reading competence, they appeal to and organize sensory perceptions.

IDEAS AND PRACTICES OF SCIENCE DISSEMINATION ACROSS THE 1949 DIVIDE

After the central government of the Nationalists settled in Nanjing in 1927, natural sciences started to receive massive institutional promotion. The state intervention was regarded as necessary to ensure the development of science and technology as essential means of national modernization. Official policies and regulations for higher education leaned toward natural sciences and engineering at the expense of the humanities and law. The state bureaucracy was to be "scientized" (*kexuehua* 科学化) to improve its efficiency, and civic service was to be professionalized. Thus, the notion of "science" was not only connected with technological modernity but also entailed the assumption of rationalizing bureaucracy and society. Founded on this notion of "science" was an elitism with "technocratic arrogance based upon unchallengeable expertise."[3] These elites took it for granted that the dissemination of modern scientific knowledge would make a new and advanced nation-state of China come true. A case in point is the China Scientization Movement mentioned in the previous chapter, which was partly subsidized by the Nationalist government.[4] The "spirit of scientization," according to Zhang Qiyun, one of the permanent board members of the Association for the China Scientization Movement, was to emphasize the application and propagation of science.[5]

The association's plan and practices of bringing science to the populace also attempted to appeal to both "scientism and cultural nationalism" and to "reconcile the relationship between science and Chinese culture."[6] Magazines and journals should be published to arouse the general interest in modern science; Western materials on science should be systematically translated; and indigenous materials should be sorted out in light of modern science. In terms of organization, the association set up branches in various provinces, which published their own popular science magazines. The Beiping branch of the association even published a pictorial for children—*Children's Science Pictorial* (*Ertong kexue huabao* 儿童科学画报). Its irregularly published newsletters (altogether seven of them) provided minutes of meetings, work reports, and forums to improve the effectiveness and efficiency of their scientization work, which included science exhibitions, radio lectures, science competitions for pupils and students, as well as night schools for workers and peasants. The China Scientization Movement ended in 1937, when the Second Sino-Japanese War broke out.[7]

The 1930s also witnessed, in Sigrid Schmalzer's words, "the first real explosion" of science dissemination efforts in Chinese society in form of cultural movements, popular science books and magazines, and science education programs.[8] This surge included, among others, the Movement of Marrying Science Down (*kexue xiajia yundong* 科学下嫁运动, c. 1931–35) organized by Tao Xingzhi 陶行知 (1891–1946), which tried to bring science to children and adults with no or low literacy. The *Children's Science Series* (*Ertong kexue congshu* 儿童科学丛书) consisting of 108 titles was published, a Children's Science Correspondence School (*Ertong kexue tongxun xuexiao* 儿童科学通讯学校) was set up, and five-minute popular science radio lectures were broadcasted for the urban illiterate.[9]

In the Communist bases in Jiangxi and then in Yan'an, science education had always been an integral part of the school curricula. In 1940 the Society of Natural Science Research of the Shan-Gan-Ning Border Base (*Shan Gan Ning bianqu ziran kexue yanjiuhui* 陕甘宁边区自然科学研究会) was founded, whose tasks were "to launch the Natural Science Popularization Movement, carry out natural science education, bring natural scientific knowledge to the masses so as to help them embark on the scientific and progressive road with new consciousness, customs, and habits."[10] This signals the beginning of the CCP's institutional efforts to disseminate scientific knowledge and thinking, which would then include lectures and exhibitions, popular science writings in the Party newspaper *Liberation Daily*, and most notably, various forms of folk culture—folk songs, local drama, and dance as well as raree shows (*la yangpian* 拉洋片)—adapted for the purpose of science dissemination.

Therefore, the Nationalists and the Communists shared similar goals of building a modern nation-state China by disseminating modern scientific knowledge, developing technology, and removing "superstitious, ignorant, and backward" thoughts and unhygienic habits of the populace. Both harbored the optimism in the positive role of science in achieving national modernization and civilizational progresses. Both tried to persuade its citizens to learn and study science in order to become more productive laborers for the nation. During the process of science dissemination, they faced the same challenge of integrating indigenous knowledge into the notion of "science," or excluding it. Yet they held different philosophical views on science and therefore attached different socio-political meanings to it. Whereas the Nationalists tried to combine science's power of rationalization and developing new technology with cultural nationalism to realize its dream of a scientific China, the CCP emphasized the materialist aspects of natural science and its association with the ideology of Marxism-Leninism. As discussed in chapter one, it perceived natural science as the knowledge of "objective reality" (*keguan xianshi* 客观现实) and promoted it to "strengthen political, economic, and cultural constructions" and "increase the welfare of the people."[11]

When the PRC was founded, the CCP predicated its interpretation of "science" on the cultural authority of science already established in Republican China and continued to make use of the hierarchical, central-branch organizational structure of the China Scientization Movement. Leftist scientists and popular science writers such as Gao Shiqi 高士其 (1905–88) and Dong Chuncai 董纯才 (1905–90), who contributed to the *Children's Science Series* in the 1930s, readily joined the PRC's efforts and became prominent figures in the field. While further consolidating the authority of "science" with its own institutionalization, the new China would add the notion of "the people" to "science."

INSTITUTIONALIZING SCIENCE DISSEMINATION IN THE PEOPLE'S REPUBLIC OF CHINA

In November 1949, the Bureau of Science Dissemination under the leadership of the Ministry of Culture of the Central People's Government (*Zhongyang renmin zhengfu wenhuabu kexue pujiju* 中央人民政府文化部科学普及局) was founded. Regarding science dissemination as a task of concerted efforts of various cultural institutions, it held seven discussion sessions (*zuotanhui* 座谈会) between December 1949 and January 1950. The minutes of these meetings were published in the *Newsletter of Science Dissemination* (*Kexue puji tongxun* 科学普及通讯, 1950), the official journal of the bureau. The variety of the attendants at these meetings shows both the importance attached to science dissemination by the bureau and its ambition to reach every level of the society. Participants of the sessions, among others, included:

science education teachers of primary and middle schools, university teachers and students, employees of research institutions and natural museums, the ministries of agriculture, education, and culture, the administrations of motion picture and cultural heritage, national trade union, national woman's federation, and the central committee of Communist Youth League.

The contents of these discussion sessions are essential to understanding the basic idea of the People's Science, because they discussed the relationships between the making of the people and the making of science. Science dissemination aimed "to propagate materialism and to enable the laboring people to grasp techniques and technology for production as well as the laws of natural development so that they are qualified for the tasks of national [economic] production and construction."[12] The attendants saw that the field of science before 1949 had been dominated by bourgeois worldviews and populated by scientists immersed in feudal-imperialist culture. Therefore, they concluded, the pre-1949 science dissemination only toyed with scientific anecdotes and catered to the taste of urbanites instead of dealing with the reality of the Chinese life. The dissemination movement in the PRC, on the contrary, would involve 90 percent of the Chinese people—especially the laboring people such as peasants, workers, and soldiers—and consider both their needs and contributions. Peasants wanted to learn modern technology to eliminate crop pests and epidemic diseases of livestock, and they presumably had also accumulated thousands of years of experience that manifested themselves in so-called indigenous recipes and methods (*tufang* 土方, *tufa* 土法), which should be interpreted, sorted out, and categorized in light of modern science.

This emphasis on local knowledge was reminiscent of the case of Chinese Medicine in Republican China in their same awareness of reconciling different knowledge systems. It, however, also harbored a dream of flattening the hierarchy of social classes: industrial workers displayed intensive interest in learning new techniques and technology to increase their productivity, which in turn would stimulate their own innovations; soldiers were included as the audience of science dissemination because they were an army of production as well as of national defense. According to the minutes the bureau and its provincial branches should promote, organize, and lead the movement by planning its contents, guiding its direction, and preparing the necessary dissemination materials such as popular science reading materials, and teaching and audio-visual materials. The use of nationalized (*minzuhua* 民族化)—and mainly non-textual—forms was advised to attract the audience, such as the Rice Sprout Song (*yangge* 秧歌) dance, bamboo clappers, folk drama, posters, lantern slideshows, motion pictures, and exhibitions. All these materials, like the book *The Inventions of Our Ancestors*, were intended to propagate indigenous scientific achievements of China to cultivate patriotism while at the same make the populace feel the relevance of scientific knowledge to their work and life. In addition, the agency of the laboring people should be initiated and their creativity should be stimulated in the science dissemination work to overcome the general shortage of resources.[13]

These discussions clarified the basic issues of who could count as "the people" and what could count as "science." Following the Marxist-Leninist theory of class struggle, peasants, workers, and soldiers made up the major part of the laboring people (*laodong renmin*). They were, on the one hand, able to offer experience-based, indigenous methods that were developed from their practices of economic production, and on the other hand, eager to learn new technologies based on modern science to solve practical problems and hence increase economic productivity. Sigrid Schmalzer observes that science dissemination in the PRC encouraged "a greater emphasis on technology than on natural science,"[14] which manifests the state's developmentalist mentality in understanding science. Reports published in the *Newsletter of Science Dissemination* and its renamed successor *Science Dissemination Work* (*Kexue puji gongzuo* 科学普及工作, 1951) repeatedly mentioned that peasants and workers were more interested in grasping new techniques/technology than in learning general scientific knowledge or scientific principles of technology.

In August 1950, the Bureau of Science Dissemination as an administrative institution became superfluous. Most of its functions—except for the management of science halls (*kexue guan* 科学馆)—were taken over by the All-China Association for the Dissemination of Scientific and Technological Knowledge (*Zhonghua quanguo kexue jishu puji xiehui* 中华全国科学技术普及协会), whose headquarters and branches were led by the Bureau of Culture and Education (*Wenjiao weiyuanju* 文教委员局).[15] The four goals of the association summarized all the CCP's concerns in disseminating "science," namely:

1. to enable the people to grasp scientific methods of production and advocate the scientization of production techniques to contribute to the economic construction of the new democracy;
2. to expound natural phenomena, scientific and technological inventions so as to dispel regressive, superstitious thoughts;
3. to propagate scientific and technological inventions of the Chinese people in order to cultivate a new patriotism; and
4. to popularize medical and hygienic knowledge to safeguard the health of the people.[16]

In September 1958, the China Association for Science and Technology (*Zhongguo kexue jishu xiehui* 中国科学技术协会, CAST) was founded, which has served as the institution responsible for science dissemination in the PRC up to today.[17]

REFORMING THE "OLD" IDEAS OF POPULAR SCIENCE

The People's Science was a complex social engineering project steered by the state, which incorporated the new state's attempt to create an economically productive, intellectually engaged, and politically progressive "(laboring) people" and its pragmatic concerns of modernizing the country with cost-effective investment of money and labor. No matter if this project was successful or not, the ideas and practices of the People's Science had drastically changed the populace's perception of scientific knowledge. For professional scientists/scientific workers, their tasks were not only tailoring the content, form, and methods of science dissemination for the masses, but also "distilling the creations and experience of the masses in the scientific way" so as to produce a "Chinese and lively" science (*Zhongguo de, huoshengsheng de* 中国的，活生生的).[18]

The first wave of endorsing the agency of the masses in knowledge production took place against the larger background of the Korean War (1950–53), when science and class politics were combined to forge patriotism, a key issue for the young PRC's state-building. In 1951, Gao Shiqi, a disabled popular science writer who had studied chemistry and medicine in the United States in the 1920s, proposed to "build a People's Science of patriotism" in *Science Dissemination Work*. He eulogized the wisdom of the Chinese people, praised the value of manual work, and portrayed a highly positive prospect of science under the political system of socialism.[19] Following this updated description of the People's Science was the journal's editorial calling for spreading the people's experiences in economic production.[20] Meanwhile, the rise of the People's Science was accompanied by the reforming of the "old," pre-1949 ideas of popular science and its producers, which destabilized the authority of professionals and undermined the notion of science as an autonomous field whose core is untouched by its political, social, and cultural contexts.

Popular science magazines, which counted over one hundred titles, were the largest legacy that the PRC inherited from Republican China. The magazine *Venus (Taibai* 太白) founded in 1934 by the educator Chen Wangdao 陈望道 (1891–1977) and a group of left-leaning writers and popular science writers, for example, celebrated science and democracy as the path to modernization. Many of its active contributors, such as Zhou Jianren 周建人 (1888–1984), Gu Junzheng 顾均正 (1902–88), and the aforementioned Gao Shiqi, would continue their careers in the PRC. Meanwhile many magazines would be taken over by the new state. *Kexue huabao* 科学画报, launched by the intellectual elites of the Science Society of China (*Zhongguo kexueshe* 中国科学社) in 1933,[21] for example, was taken over in 1953 by the Shanghai branch of the All-China Association for the Dissemination of Scientific and Technological Knowledge. Its original English title *Popular Science Monthly*, which may salute the American magazine *The Popular Science Monthly* founded in 1872, stopped appearing on the cover of the magazine after the takeover. Another example is *Kexue dazhong* 科学大众 re-launched and chief-edited in 1946 by Wang Tianyi 王天一 (1916–2002), who served as one of the editors when an early version of the magazine briefly appeared in 1937. Its English title *Scientific China Monthly* also resonated with an American popular science magazine—*Scientific American* founded in 1845. Wang's magazine was taken over by the Commercial Press in 1950 and would become one of the most subscribed popular science magazines in the first decade of the PRC.

Print media such as popular magazines and newspaper supplements might not be the most effective means of science dissemination at the time, but they were considered as potentially effective ones given the nationwide literacy movement in the early PRC. Many of the magazines launched before 1949 had to adjust their guidelines and styles to contribute to the new, socialist culture. This may mean public self-criticisms, as the case of *Kexue dazhong* in early 1950 shows. In order to complete its transition to a proper transmitter of scientific knowledge and thinking, the magazine confessed a series of its problems, including the influence from American popular science magazines, its lack of political consciousness and "scientific"—that is, dialectical materialistic—view, and its contents not related close enough to the life of Chinese masses. The editors, following closely the CCP's definition of its socialist culture and education, avowed to correct their mistakes by making the magazine national, scientific, and for the masses.[22]

The "Sinicization of science" (*kexue Zhongguohua* 科学中国化) proposed by Gu Chaohao 谷超豪 (1926–2012), a mathematician and a cadre responsible for the science dissemination campaign in Zhejiang Province, promoted the mutual learning of the intellectuals/professionals and the masses/nonprofessionals. Experience-based knowledge produced by the masses, stated Gu, should be improved through the campaign, which, in turn, equipped the masses with the general theories of modern science and stimulated their creativity. In the process, the intellectuals would also receive their education from the wisdom and experience of the masses.[23] Gu's idea affirmed a reciprocal relationship between theory and first-hand experience in knowledge production and implied the possibility of class-leveling.

Toward the end of 1950, especially after China entered the Korean War, science dissemination was explicitly tied to developing patriotism, supporting the state's economic production plan and policy, and introducing the Soviet scientific achievements to showcase the superiority of socialism.[24] The *Newsletter of Science Dissemination* reported that faculty members and students at universities held science lectures and exhibitions in their departments for the general audience. University students devoted themselves to working as volunteers in night schools.[25] Professional scientists were mobilized to participate, to reform themselves, and to demonstrate the usefulness of their knowledge. The mathematician Hua Luogeng 华罗庚 (1910–85) and his students are a case in point. Hua published articles in the *People's Daily* and popular magazines to express his gratitude to the new regime for understanding the importance of mathematics for building a new society.[26] In 1958 he became a household name for his efforts of popularizing the critical path methods (CPM, *tongchoufa* 统筹法) to organize and manage economic production. In the 1960s and 70s he took his students to visit several hundred cities and thousands of factories in more than twenty provinces, where they gave lectures on CPM and optimization methods (*youxuanfa* 优选法) to workers and technicians. These efforts were believed to have inspired tens of thousands of technological/technical improvements and innovations, which assumingly produced immense profit.[27]

Central to the reforming of the "old" popular science in the early 1950s was the emphasis on the agency and needs of the laboring people in knowledge production and dissemination. Whereas the public discourse of science may have opened up to a more inclusive way of knowledge production, many uncertainties—for example, the absence and/or ambiguity of the authority or the diverse epistemologies of the populace—lurked in the "new" ways of knowledge production and transmission, making them highly experimental and full of risks. These constantly produced tensions and confusions in the dissemination campaign, a fact intensified by the fluctuations of the state's political, economic, and education policies. We will demonstrate this point in the case studies. For the rest of this chapter we focus on how non-textual means of science dissemination presented the People's Science, presuming that they may have attracted larger audiences because they required less literacy to understand. We will focus on three visual-based forms—exhibition, illustration, and film—to analyze the ambitious and ambiguous role they play in conveying the idea of the People's Science through "the figure (*xingxiang* 形象) of objective facts."[28]

EXHIBITING SCIENCE

A 1958 poem titled "Let's Go Visit the Exhibition" (*Kan zhanlan qu* 看展览去) in the *People's Daily* expressed the joy of visiting an exhibition of national industrial transport: "I am so happy/When I enter the exhibition hall." The first-person perspective suggests the first-hand experience and the spectator sees the miniature model demonstrating the new, dense network of national industrial transport lines, which s/he metaphorically phrased as "tens of thousands of flowers." The result of visiting the exhibition, as claimed by the poem, was the spectator's conviction that industrialization would be realized soon.[29] Thus, the tangible artifacts conveyed abstract ideas, such as the superiority of socialism and accelerated modernization. Making material artifacts a central part of exhibitions not only strived to present science in a lively fashion, but perhaps also intended to follow the Maoist dialectical materialism that perceives the perceptual stage of knowledge about the material world as the necessary step preceding the rational stage.

The exhibitions in the 1950s and 60s exploited their role of the persuader and the instructor, a traditional role which is very different from that in contemporary museum theories, which seeks to stimulate the spectator's agency to question the authority of displayed artifacts.[30] Artifacts on display were selected, made, and organized in relation to the space they occupied in such a way that their spectators would form certain opinions. In this sense, artifacts displayed on exhibitions of science and technology in the early PRC were anything but neutral (arti)facts. Instead they functioned as visual and tangible evidence to persuade and to shape public discourse and opinions about scientific knowledge and its taxonomy. They informed the spectator on what counted as science and helped them to form their sense of reality and order the world in relation to their own lives.

It can be assumed that exhibitions were fairly attractive because they provided spectacles of the curious and the new, but this also brought up the question of their effectiveness in conveying the message. The 1950 grand exhibition of scientific knowledge in Beijing, for example, took place during the Spring Festival when people were used to enjoy themselves on folk fairs. Consisting of three major themes—hygiene for women and children, evolution from ape to man, and general science, this exhibition was propagated with fanfare and boasted an audience of 100,000 in twelve days, with a variety of workers, peasants, soldiers, students, urbanites, housewives, etc.[31] In the countryside, science dissemination workers made a point to hold small-scale exhibitions on folk festive occasions such as temple festivals (*miaohui* 庙会), livestock shows (*luo ma dahui* 骡马大会), or folk fairs (*ganji* 赶集). These exhibitions laid their focuses on everyday science, such as new agricultural techniques and personal hygiene. An exhibition in Pingyuan Province (covering parts of today's Henan and Hebei), for example, staged living newspapers (*huobaoju* 活报剧) to attract the audience in order to propagate modern hygienic knowledge and denounce shamanist healing methods.[32]

According to these reports, science exhibitions on folk festive occasions were an economical and effective way of attracting the rural audience who lived scattered in the countryside. This statement assumed that the audience with minimal or no literacy would readily accept the intended message of (the miracles of) science and "scientific" materialism that replaced the original religious message of folk festivals, but this assumption failed to consider the cognitive gap between the exhibition workers and their rural spectators, who did not necessarily share the same "knowledge base" and "intellectual paradigm." The result is that the spectators could, or would, not interpret what they saw in the way expected by the exhibition workers, a point elaborated by Miriam Gross in her case study of the rural participant's experience with the microscope.[33] For the visitors of science exhibitions on folk festivals, therefore, chances are that they would interpret the exhibited materials within their epistemological frameworks—as a new form of entertainment or even for erotic purposes. In the end of 1951, for example, science dissemination workers complained that some visitors toured the exhibition perfunctorily (*zouma guanhua* 走马观花), or that they came with "vulgar" interests.[34]

The epistemological and, consequently, pedagogical chaos generated on folk festive occasions were minimized in the (often permanent) exhibitions held in public yet closed—hence more controllable—spaces such as the science hall. The history of the science hall can be traced back to 1895, when the Society for the Study of Self-strengthening (*Qiangxuehui* 强学会) in Shanghai decided to set up a science museum by arguing for the explanatory power of pictures and artifacts: "Words can explain; where their meanings are unclear, pictures have to be used to illustrate. Where the shapes of pictures fail to illustrate, artifacts have to be used to demonstrate."[35] In 1928, chemist and educator Wang Jin 王琎 (1888–1966) proposed on the national education conference to establish a science hall in every province.[36] The first science halls were founded in Fujian and Shanxi Provinces, with comprehensive goals of "popularizing scientific knowledge to the people, aiding school science education as well as promoting researches in natural sciences and their applications." The idea of the science hall had received serious consideration by the Ministry of Education of Republican China since 1941.[37]

By 1949, fifteen science halls had been established in various provinces and in the city of Shanghai. They held permanent and traveling exhibitions, invited experts to give lectures and aired short radio lectures, brought film screening to the countryside, provided laboratories for school subjects of physics, chemistry, and biology, manufactured instruments and posters for primary and middle school science education, and financed science competitions. In the early 1950s the Central People's Science Hall (*Zhongyang renmin kexueguan* 中央人民科学馆) was built in Beijing, which served to exemplify the idea of the People's Science: in addition to the functions mentioned above, it also propagated the Chinese people's inventions and new ideas of production so that "science can take root in the soil of the new China, germinate, and bloom."[38]

In December 1950, an exhibition titled "The Northeast and Korea" was mounted in the People's Central Science Hall. Boasting an audience of up to 8,000 per day, the exhibition also published its materials in the book form—as an atlas, in which most of its visual and textual materials were collected.[39] Combining the information in this book with a contemporary descriptive report on the exhibition,[40] we try to reconstruct a tour of the exhibition. More precisely, we show how geographical knowledge about the Northeast and Korea was presented as "objective facts" and organized spatially to argue for the state's political and military decision to join the Korean War, or as it was called in the Chinese media, the War to Resist America and Aid Korea.

The exhibition consists of three parts. The first part argues, with maps and statistic figures, that the attempt of "American imperialists" to build its hegemony over the world is doomed. The second part employs artifacts—specimens, models, replicas, etc.—to demonstrate the strategic importance of the Northeast for the new China's economic development. The last part presents photos and maps to introduce Korea as a geopolitically strategic partner and an ideological ally of the PRC while denouncing "American imperialists" for invading the peninsula. As the spectator, we first encounter a large globe (one meter in diameter) at the entrance of the exhibition, which visualizes the distribution of the two political camps. Following the Soviet Union, the only socialist country, are the "new democratic countries," that is China, Mongolia, the Democratic People's Republic of Korea, as well as the countries in East Europe. Peoples rising up against imperialism, indicated by little figures holding guns in Malaysia, Indonesia, and Greece, etc., are counted as the growing influence of the socialist camp. The rest of the world is the "followers" and "slaves" of the "imperialist camp led by the US." Although the visuals seem to indicate more political influence of the United States, the caption nevertheless predicts that the "camp of peace and democracy led by the Soviet Union" is becoming stronger everyday while that led by American imperialists is declining. This prediction is backed up by a map and a table including undefined "left-leaning" people and those "advocating peace and democracy," which more or less cover the whole world, as forces against "American imperialists." The exhibition goes on to deploy a map to show "The Futile Fantasy of American Imperialists: Besieging China (*Meidi de huanxiang: baowei Zhongguo* 美帝的幻想：包围中国)."[41]

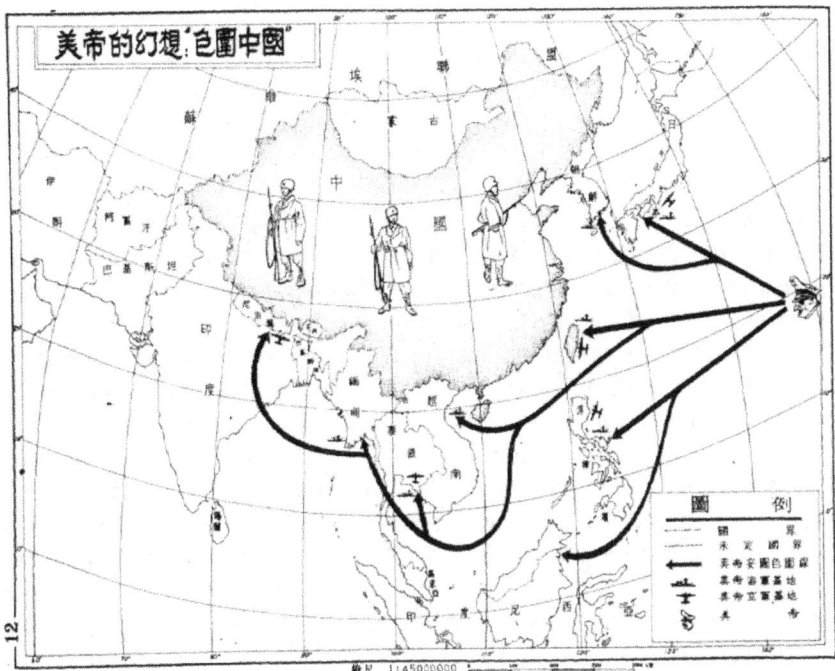

Figure 2.1. Zhongyang renmin kexueguan 中央人民科学馆. *Kangmei yuanchao yundong zhong de dongbei yu chaoxian tuji* 抗美援朝运动中的东北与朝鲜图集 (Atlas of the Northeast and Korea in the Campaign of Resisting America and Aiding Korea). Shanghai: Zhonghua shuju, 1951.

The map legend indicates that the American imperialist is represented by the caricature of Douglas MacArthur (1880–1964), the Commander-in-Chief of the United Nations Command. From this caricatured head shoot out three bold arrows over the Pacific, indicating the major paths that the American imperialist will take to besiege China. One of them is directed to Japan which furcates into Korea. Others aim at American navy and air force bases in South and East Asia. Mainland China, the desired object of the imperialist, is defined not just by its borders and territory, but also by three soldier figures vigilantly facing three directions, except for that of Mongolia and the Soviet Union. The one facing Korea levels up his gun and appears more on alert than the other two. The fact that the map, a visual means of conveying geographical knowledge, exudes such thinly veiled hostility against the United States shows that the exhibition expected to affect the spectators' emotion rather than their intellectual faculties.

Assured that the American imperialist and his attempt to besiege China are doomed, the spectator enters the second part of the exhibition, which shows the strategic importance of the provinces in the Northeast. Maps illustrate the rich natural resources in agriculture, forestry, and its developed industry and transportation infrastructure as well as the area's capacity to accommodate migrants. They are, in turn, supported by a selection of specimens of agricultural and textile products, furs, wood, minerals, miniature replicas of blast furnaces from Anshan Iron and Steel Works (the AnSteel, *Angang* 鞍钢), and a model of the Fengman hydroelectric power station. The fact that the AnSteel, the power station, and the Northeast's major transportation infrastructure were built during its time of colonization by Japan were not mentioned. These maps and artifacts on exhibition thus intended to convince the spectator of the current productivity and future potential of the Northeast, which should lead them to conclude that this area must be protected to ensure the security and further development of China.

The third part exhibits photos and maps. The former portray the beautiful landscape of Korea, the sufferings of the Korean people, as well as the "waking-up" American prisoners of war protesting the invasion of Korea. The latter visualizes the geographical position of Korea in order to argue that Korea can be used as the springboard by the United States to grab the Northeast of China. In contrast to the first part that intends to arouse hatred of the United States, this part attempts to cultivate in the spectator a feeling of camaraderie toward the Korean people.

If the spectator follows the three parts of the exhibition as intended by its designer, then the knowledge and emotion s/he develops would correspond to and be reinforced by the slogans written in large characters on the wall of the exit: "To liberate Korea/To safeguard our Northeast/To realize the independence of China/To achieve the victory of the peaceful and democratic camp/We have no choice/But resolutely resist America and aid Korea."[42]

With the benefit of hindsight, we know the information that this exhibition conveyed—"the American imperialists' invasion of Korea"—was not true. Then how did this exhibition argue for the necessity—if not the inevitability—of the state's decision to join the Korean War in the framework of science dissemination? The tour of the imagined spectator described above shows that the new state resorted to scientific presentation (maps, statistics) and material evidence (photos, models, specimen) to stir up emotion and endorse its argument as true. The alliance of the exhibition with "science" was further reinforced by its location in the science hall, an authoritative space of science education. In the short preface of the book, the editors said that its target readers were those with higher primary school education, that is, someone with about four to five years of schooling. They were very possibly also the prospective spectator, for whom the exhibition was designed. For them, persuasion by means of sensory perception would work more effectively than that through lengthy texts. To sum up, this exhibition borrows the presentational methods and cultural authority of an autonomous notion of "science" to stir up patriotic emotions and to form the opinion of a general public (with relatively low literacy) on a political decision.

Science halls as independent *danwei* 单位 (work unit), however, did not last long in the PRC, perhaps because it was not a cost-effective means of science dissemination and its influence was limited by time and space. Most of them would have been restructured into museums by the mid-1950s. In 1951 the People's Central Science Hall and all its employees were transferred to the Beijing Museum of Natural History, which was then in preparation. Between 1954 and 1956, this museum held several grand exhibitions on topics such as mineral resources in China, agriculture, the liberation of Taiwan, and the flood control of the Yellow River.

SCREENING SCIENCE

Science education films (*kexue jiaoyu dianying* 科学教育电影, or briefly *kejiao pian* 科教片) has been a distinct film genre in the PRC to disseminate scientific knowledge. Film technologies not only offer moving images accompanied by the voiceover explanation, but are also able to display what is invisible to the naked eyes and use specific techniques such as montage to create a strong narrative and argument.

The Commercial Press in Shanghai, which played an essential role in developing science education textbooks in early twentieth-century China,[43] also pioneered the use of film for education. Starting with used facilities bought from an American, the Commercial Press made twenty "education films" between 1918 and 1923. The Association of China's Education Film (*Zhongguo jiaoyu dianying xiehui* 中国教育电影协会) was founded in July 1932. Chen Lifu 陈立夫 (1900–2001), one of its organizers who was also one of the founders of the China Scientization Movement, rejoiced in the film's function for education and culture, in particular popularizing scientific knowledge. The School of Science at Jinling University in Nanjing produced science films around 1936, and China's Education Film Studio was set up in wartime Chongqing in 1942.[44] Communist filmmakers made their first work *Preventing Bubonic Plague* (*Yufang shuyi* 预防鼠疫) as early as 1948 with the equipment and technicians from the former Manchuria Motion Company (*Manying* 满映) studio in Changchun.[45]

Before the first science education film studio of the PRC was established in February 1953 in Shanghai, however, it was self-made lantern slideshows, so-called "homemade film" (*tu dianying* 土电影) that were actively promoted in the early 1950s as an inexpensive, simple, and flexible way of science popularization. Slideshow facilities were relatively easy to carry and may use gas lighting when electricity was unavailable, which made these shows mobile enough for traveling in the countryside. Even more important was its adaptability to local needs: its contents could be erased and written anew to speak to local people and their issues. It was considered as an effective tool of science dissemination on local festive occasions.[46] In 1951, the Manufacturer of the Audio-Visual Education Instrument (*Dianhua jiaoyu gongju zhizaosuo* 电化教育工具制造所) affiliated to the Science Dissemination Bureau produced more than thirty titles of slideshow.[47] Another major producer was the Agriculture Film Society (*Nongye dianyingshe* 农业电影社) affiliated to the Ministry of Agriculture, which focused on new technologies and high yield experiences.[48]

In December 1953, the guidelines for science education film were set up by the State Council in a directive to invigorate filmmaking. The directive, following closely the general guidelines of science dissemination, required that "science education films use the materialist worldview to explain natural and social phenomena, and to promote scientific and technical knowledge which is related to the masses in their daily life, economic production, and which suits their knowledge level." The fact that the State Council expected fifteen science education films every year in the next four years shows that this filmic genre gained more attention than other forms of documentary.[49]

Soon an exhibition of short science education films was announced to take place in six large cities—Beijing, Shanghai, Shenyang, Hankou, Chongqing, Xi'an—in April and May 1954, whose thirteen titles ranged from general science topics such as *Physical Exercises and Health* (*Tiyu yu jiankang* 体育与健康), *Electricity Safety at Home* (*Jiating anquan yongdian* 家庭安全用电), *Solar and Lunar Eclipses* (*Rishi yu yueshi* 日食与月食) to agricultural production such as *Water and Soil Conservation* (*Shuitu baochi* 水土保持), *Eliminating Locusts* (*Xiaomie feihuang* 消灭飞蝗), and technical innovations such as the famous *Hao Jianxiu Work Methods* (*Hao Jianxiu gongzuofa* 郝建秀工作法) in the textile industry.[50] After a surge of science education films in the Great Leap Forward years, the Central Propaganda Department issued guidelines in July 1961, which required that the genre focus on introducing modern science and technology to the masses and stop treating social science topics.[51] Their key points were summarized in 1962 as follows: the audience should "be able to understand, learn, and use" (*kan de dong, xue de hui, yong de shang* 看得懂，学得会，用得上) the knowledge conveyed in the films.[52] The table provided by Matthew Johnson in his study on science education film in modern China shows that the Beijing Science Education Studio produced eighty-nine titles of science education films between 1960 and 1964, among which those of agricultural themes amounted to thirty-one titles, indicating the importance laid on agricultural science and technologies at the time.[53]

On the 1963 exhibition of science education films during the National Science Education Film Propaganda Week in ten cities, films such as *The Jewel Wasp and the Pink Bollworm* (*Jinxiaofeng yu honglingchong* 金小蜂与红铃虫, 1963) and *Clever Planting and More Harvest* (*Qiao zhong duo shou* 巧种多收, 1962) were praised for their promotion of technical innovations in China's pursuit of agricultural modernization.[54] Known as one of the best science education films made in the Mao Era, *The Jewel Wasp and the Pink Bollworm* (directed by Han Wei 韩韦, photography by Xu Leding 徐乐定 and Zhu Yonggao 朱永镐) introduces a method of biological pest control—to use the jewel wasp to kill the pink bollworm, a pest in cotton boll. This twenty-minute black and white science education film has been celebrated for its skillful application of film technologies (especially cinematography) to constructing a dramatic narrative.[55]

The story employs the friend-foe rhetoric to describe the relationship between the jewel wasp and the pink bollworm as the struggle between good and evil. The pink bollworm is identified as an enemy, because it damages the cotton boll, "the fruit of our labor." Researchers in Hubei Province have spent three years (1955–58) looking for an economical and effective solution and at last found the jewel wasp, "our insect friend." With cinematographic facilities, the film is able to offer its audience a magnified view of the jewel wasp–killing the bollworm, dramatized by the use of montage. The fight between the two insects, staged as a fight between good and evil, is magnified twenty to fifty times in order to present a view that is usually too tiny for naked eyes. The cutting and editing of the shots create a tension in the narrative to emphasize the jewel wasp's hard struggle to achieve the victory.[56]

This film also exemplifies the People's Science in the 1960s, which emphasized concerted efforts of researchers/technicians and peasants as well as the methods of experiment and local adaptation. As the story goes, the researchers have carried out experiments repeatedly and then successfully made the jewel wasp reproduce in the incubator of their laboratory. With a hypothetical estimation that this biotechnological method can reduce 70 percent of the damage, they send out the young wasps to the people's communes for the "real struggle." The peasants in a commune, represented in diegesis by a young man and a young woman, adapt this method to their own situation. In the absence of the incubator, they create the desired temperature with the oil lamp to make jewel wasps reproduce. Then they place the wasp wherever they deem proper. In the end researchers and peasants come together to check the result and find the method highly effective. Their concerted efforts embody a "standpoint epistemology," which, according to Sigrid Schmalzer, posits that

Figure 2.2. A magnified view showing the jewel wasp fighting against the pink bollworm, still frame of the movie *The Jewel Wasp and the Pink Bollworm* (*Jinxiaofeng yu honglingchong* 金小蜂与红铃虫, 1963), minute 7:43.

"people contribute differently to the production of knowledge based on their social position." The fact that the peasants are portrayed as young symbolizes their "newness"; they should be the generation educated and empowered by the People's Science—the educated "peasant technicians."[57]

The reception of this film at the time, especially by the rural audience it intended to address, remains difficult to fathom. According to a report of the *People's Daily*, a female peasant from Shanghai said that they used to get chicken to eat up the pink bollworm, but the jewel wasp seemed more effective.[58] This statement does not sound very convincing because, as the reporter admitted, she may have never seen a jewel wasp in her region. Matthew Johnson quotes from 1963 conference notes in the Beijing Municipal Archive to argue that screenings in rural areas worked better when the audience saw that the film contained knowledge that could produce tangible results.[59] Other secondary materials show that science education films—their voiceover explanation, images, narrative sequence, etc.—were by no means self-evident for rural viewers. For example, a film projection team for the rural audience in Jiangsu Province had to insert their own explanation to facilitate the understanding of the viewers, which mean, they had to learn the content of the film by heart and understand well the perspective and vocabulary of the local viewers.[60] Another team in Guangdong Province sought help from local peasants to rewrite the voiceover script so as to attract more audience.[61] These two instances indicate that most science education films, despite their themes on agriculture, may not be so well-tailored for the rural audience after all.

Therefore, the cognitive gap also appeared in the screening of science education films. Seeing is certainly not believing, even when the visual images are explained. In the early 1980s the "propaganda before the movie" (*yingqian xuanchuan* 映前宣传) was still needed to help the rural audience to grasp as much as possible the instructional contents from the movie.[62] A 1982 article published in *Popularization of Film* (*Dianying puji* 电影普及) is revealing with a blunt passing remark: "In the past, when we released science education films, . . . there were the problems that . . . the projector was not willing to show it and the masses did not want to watch it."[63] This remark raises many questions about the effectiveness of education film screenings for the—perhaps not just rural—audience. In comparison with their rural compatriots, urban residents, especially industrial workers and students, enjoyed better educational opportunities and economic conditions as well as more accessibility to public facilities such as public libraries and science halls. For them, popular science magazines were a more common access to scientific knowledge.

ILLUSTRATING SCIENCE

Popular science magazines contributed to making the People's Science by visualizing an optimistic future of modern China empowered by science and technology and shaping the consciousness of the working class as the knowledge producer. These are amply evidenced in the magazines with the top three largest subscription in the Mao era—the abovementioned *Science Pictorial* (*Kexue huabao* 科学画报, 1933–66), *Popular Science* (*Kexue dazhong* 科学大众, 1946–66), and *Knowledge Is Power* (*Zhishi jiushi liliang* 知识就是力量, 1956–62). In the following, we examine visual and textual contents in *Knowledge Is Power* and *Science Pictorial* in order to show their (complementary) roles in envisioning the People's Science. *Science Pictorial* interpreted and propagated the official policies concerning public health, technological innovations, and building the Chinese genealogy of science, while *Knowledge Is Power*, following the style and ideas of its Soviet namesake Знание-сила, presented the vision of "tomorrow"—the near future—for young workers and students. Starting its publication in 1956, when the state called for "marching towards science," it celebrated the vision(s) of the socialist science by picturing the amazing ability of science to bring about social and economic progress.

The first five issues edited by the Soviet side set the example of cultivating the ethos of the socialist future among its readers. In the inaugural issue, the Russian chemist N. D. Zelinsky (1861–1953) was cited as saying that "our present is the future of our foreign friends,"[64] a quote that finds its Chinese equivalence as "today's Soviet Union is like our tomorrow" (*Sulian de jintian shi women de mingtian* 苏联的今天是我们的明天). Both dictums temporalize the spatially located Soviet Union into the future and thereby put it on a higher stage of the progression of history. From the sixth issue on, the All-China Association for the Dissemination of Scientific and Technological Knowledge and the Ministry of Labor took over editorial responsibility. The central committee of the Communist Youth League joined the editorship in 1957, indicating the magazine's task to shape future generations. The popular science writings, reports on new technology and their *potential* applications as well as Soviet science fiction in this magazine reduced the distance between "experience and expectation" by "celebrating tomorrow today."[65] As the result, this magazine offered the reader "visions of the *planned* transformation of society by rational, scientific means."[66]

It should be noted that states on both sides of the Iron Curtain attempted to demonstrate the positive role of science and technology for humanity and thus to argue the superiority of their political systems. Whereas the West claimed "an indissoluble link between scientific genius and liberal democracy,"[67] the East celebrated their scientific and technological progress under guidance of the state's central and scientific planning and the creativity of the emancipated labor force. The idea of a near, utopian future of communism is conveyed and concretized through a large number of visuals such as colored covers and spreads, illustrations, and photos in *Knowledge Is Power*. Even the fact that this journal could afford to print visuals on relatively high-quality paper was itself a rare case at the time and thus appeared futuristic as an indication of material abundance. The material achievements of Soviet science were presented in the magazine through visuals as political symbols, intending to convince its target readers—young Chinese workers—that a progressive, communist future is realizable through their efforts. Yet sometimes pictures, supposedly expressing much more than words, may betray message without intending to do so.

A case in point is a 1957 cover image portraying a large shaft-drilling rig made in the Soviet Union—the type UKB-3.6.[68] A new invention for the coal industry with a drilling diameter of 3.6 meters, the UKB-3.6 was put into operation in 1956. Using coring techniques, this shaft drilling rig made it possible "for the first time to drill such a large hole in unstable rock by the 'full-face' drilling method." It was awarded the gold medal for construction at the 1958 World Exhibition in Brussels.[69]

This cover image comes without any textual explanation in the magazine, which perhaps suggests the self-evident reputation of the machine among

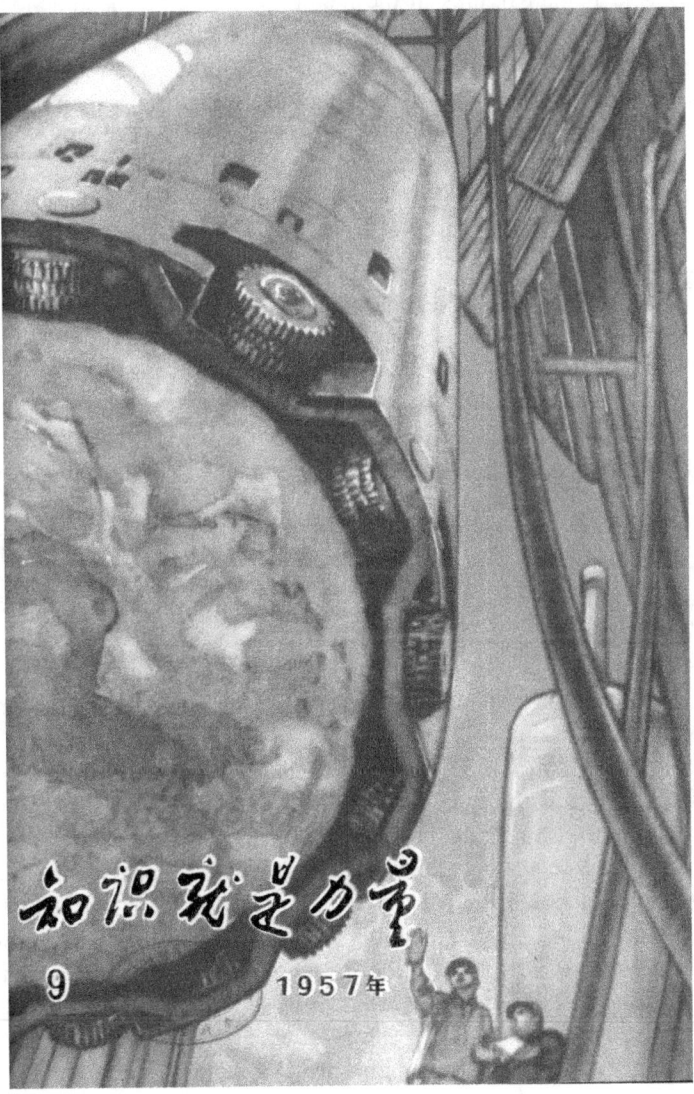

Figure 2.3. Cover page, *Knowledge Is Power* (Zhishi jiushi liliang 知识就是力量), no. 9 (1957).

contemporary industrial workers. Adopting a perspective from beneath the drill rig, the image devotes two-thirds of its space to the machine to accentuate its size. The immense power of the machine is expressed through its imposing position in relation to human figures (engineers or workers), who are squeezed to the right lower corner between this sinking monster of the machine and huge cables and platforms. The unbalanced composition of the machine and human figures in the picture visualizes a machine-man relation that does not really fit into the Marxist-Leninist theory of emancipated laborer: the power of the machine to conquer nature and to realize industrial modernization is foregrounded to such an extent that its human creator and user—or humanity in general—seems insignificant. This picture, representing the "present" technological sophistication of the Soviet Union, promised an industrialized future for the Chinese reader, who, while overwhelmed in awe by the power of the machine, may possibly forget it was a creation of human hands.

Different from the impressive visuals afforded by *Knowledge Is Power*, *Science Pictorial* engaged its readers with its interactive rubrics. From the very beginning in 1933, the section "Postbox" (*Xinxiang* 信箱) was set up to answer the reader's questions and respond to their criticism, citing both Chinese- and English-language sources,[70] a practice that disappeared when the magazine changed hands in 1953. The renamed "Brief Q & A" (*Wenti jianda* 问题简答) assumed an authoritative tone, limiting the questions to general science in relation to economic production and daily life. It explicitly declined questions on professional scientific research and any form of disease diagnosis.[71] A significant change took place in 1957, when this rubric was converted into a forum, with its new name "Let's Discuss" (*Dajia tan* 大家谈). This section not only answered readers' questions,[72] but also told stories about innovators and scientists in ancient China[73] and published reports on extraordinary natural phenomena and their explanations by readers.[74]

"Let's Discuss" in the July issue of 1957 demonstrated various ways of how the reader could participate in the forum and challenge the authority of the (editors of the) magazine. One reader, for instance, contended that diving was nothing new in ancient China—after reading in the magazine about divers and their equipment in Western sources—and evidenced his contention with a picture portraying pearl hunting in the *Exploitation of the Works of Nature* (*Tiangong kaiwu* 天工开物, 1637). Another reader argued that bamboo did not grow higher with age, as claimed by an article printed earlier in the magazine.[75] In the 1960s some readers' letters appeared academic, such as a letter taking issue with an article published in the magazine stating that goats would hardly get tuberculosis. This reader argued his/her case citing from both academic literature and his/her practical experiences.[76] Different from the Q&A rubric, these forms of discussion seemingly offered the reader—presumably the masses—opportunities of participation: the reader may supplement information, share their understanding of science and scientists, and cite their practical experiences as valid scientific evidence. All these seem to exemplify the model of knowledge production and dissemination through the interaction between professionals/experts and the masses.

Knowledge Is Power hardly had such direct interaction with its readers until 1962, a section called "Our Postbox" (*Women de xinxiang* 我们的信箱) appeared to deal with readers' discussion of scientific problems and corrections of printing mistakes. This change is indicative of the magazine's moving away from the Soviet model to encourage its reader to use their scientific knowledge, especially empirical experiences, to solve practical problems.

The People's Science developed in the first decades of the PRC was embedded in Marxist-Leninist dialectical and historical materialism, identity politics (patriotism), and class politics (class struggle) as well as a vision of communist future. It was key to the young PRC's state-building. The next three case studies on agricultural mechanization, metallurgy, and Chinese veterinary medicine shall discuss the role of various actors in seeking alternative, pragmatic ways of producing and disseminating knowledge as the People's Science.

NOTES

1. The term appeared in, for example, in Kexue pujiju fudaochu diaocha yanjiu ke 科学普及局辅导处调查研究科, "Quanguo kexueguan shiye gaikuang ji gaijin de yijian 全国科学馆事业概况及改进的意见 (An overview of science halls facilities nationwide and suggestions for improvement)," *Kexue puji tongxun*, no. 6 (1950): 106–8; and Gao Shiqi 高士其, "Jianshe aiguozhuyi de renmin kexue 建设爱国主义的人民科学 (Building a people's science of patriotism)," *Kexue puji gongzuo*, no. 2 (1951): 29.

2. Magazine titles in the 1930s and 1940s demonstrate the centrality and continuity of popular science: *Popular Science Quarterly* (*Dazhong kexue jikan* 大众科学季刊) in Sichuan between 1934–1936, and *Popular Science Monthly* (*Dazhong kexue yuekan* 大众科学月刊) in Shanghai in 1938. The journal *Scientific China Monthly* (*Kexue dazhong* 科学大众) was first launched in 1937 in Shanghai and was taken over by the Commercial Press in 1950. It should be noted that the popular science lectures held by the People's Central Science Hall in 1950 were also entitled *Dazhong kexue*.

3. William Kirby, "Engineering China: Birth of the Developmental State, 1928–1937," in *Becoming Chinese: Passages to Modernity and Beyond*, ed. Wen-hsin Yeh (Berkeley: University of California Press, 2000), 151. See also Fröhlich, *Staatsdenken im China der Republikzeit*; Kwok, *Scientism in Chinese Thought 1900–1950*.

4. Peng Guanghua 彭光华, "Zhongguo kexuehua yundong xiehui de chuangjian, huodong jiqi lishi diwei 中国科学化运动协会的创建、活动及其历史地位 (Foundation, Activities and Historical Assessment of the Association for China Scientization Movement)," *Zhongguo keji shiliao* 中国科技史料 13, no. 1 (1992): 60–72.

5. Zhang Qiyun 张其昀, "Kexue yu kexuehua 科学与科学化 (Science and scientization)," *Kexue de Zhongguo* 科学的中国 1, no. 1 (1933): 4–11, quote 4.

6. Lei, *Neither Donkey nor Horse,* 142, 148.

7. Peng Guanghua, "Zhongguo kexuehua yundong xiehui de chuangjian, huodong jiqi lishi diwei," 66–67; Chen Hongjie, "Zhongguo jindai kepu jiaoyu: shetuan, changguan he jishu," 21–22.

8. Schmalzer, *The People's Peking Man,* 29.

9. Hong Xiangsheng 洪祥生, "Ba Zhonghua minzu zaojiu wei yige kexue de minzu: Xi Tao Xingzhi de 'kexue xiajia' 把中华民族造就为一个科学的民族：析陶行知的'科学下嫁' (Making the Chinese nation into a scientific nation: an analysis of Tao Xingzhi's 'marrying science down' movement)," *Anhui jiaoyu xueyuan xuebao*, no. 1 (1988): 79–81. See also Tao Xingzhi 陶行知, *Shenghuo jiaoyu wenxuan* 生活教育文选 (Selected Works on Life Education), ed. Hu Xiaofeng 胡晓风 (Chengdu: Sichuan jiaoyu, 1988).

10. Wang Heng 王恒, "Yan'an de kepu huodong 延安的科普活动 (Activities of science popularization in Yan'an)," *Zhishi jiushi liliang*, no. 7 (2011): 20.

11. "Jiangli ziyou yanjiu 奖励自由研究 (Rewarding free research)," *Jiefang ribao*, June 7, 1941, and "Tichang ziran kexue 提倡自然科学 (Promoting natural science)," *Jiefang ribao*, June 12, 1941. Both are collected in Zhongyang jiaoyu kexue yanjiusuo 中央教育科學研究所, *Lao jiefangqu jiaoyu ziliao—kangri zhanzheng shiqi (xia ce)* 老解放區教育資料－抗日戰爭時期 (下冊) (Materials on education of the liberated bases—the war of resistance against Japan period, Vol. 2) (Beijing: Jiaoyu kexue, 1986), 8–13.

12. "Kexue puji wenti zuotanhui zongjie 科学普及问题座谈会总结 (Summary of the forum on issues of science dissemination)," *Kexue puji tongxun*, no. 1 (1950): 3, 7 and no. 2 (1950): 20–22. A slightly different formulation is the following: "the propagation of natural scientific knowledge plays an extremely important role in cultivating the Communist worldview and enables ordinary workers, peasants, and soldiers to learn science and technology required by economic production and thereby grasp the laws of natural evolution." See "Sige yue lai de kexue pujiju 四个月来的科学普及局 (Science dissemination bureau in the past four months)," *Kexue puji tongxun*, no. 1 (1950): 2.

13. "Kexue puji wenti zuotanhui zongjie (Summary of the forum on issues of science dissemination)," *Kexue puji tongxun*, no. 2 (1950): 20–21.

14. Schmalzer, *The People's Peking Man,* 295.

15. Yuan Hanqing 袁翰青, "Kexue puji gongzuo de xin jieduan 科学普及工作的新阶段 (A new phase of science dissemination work)," *Kexue puji gongzuo*, no. 10 (1951): 201.

16. Liang Xi 梁希, "Zhonghua quanguo kexue jishu puji xiehui de renwu 中华全国科学技术普及协会的任务 (The Tasks of All-China Association for the Dis-

semination of Scientific and Technological Knowledge)," *Kexue puji tongxun*, no. 7 (1950): 129.

17. Schmalzer, *The People's Peking Man*, 68.

18. Kexue pujiju fudaochu diaocha yanjiu ke 科学普及局辅导处调查研究科, "Quanguo kexueguan shiye gaikuang ji gaijin de yijian 全国科学馆事业概况及改进的意见 (An overview of science hall facilities nationwide and suggestions for improvement)," *Kexue puji tongxun*, no. 6 (1950): 106–8.

19. Gao Shiqi 高士其, "Jianshe aiguozhuyi de renmin kexue 建设爱国主义的人民科学 (Building a people's science of patriotism)," *Kexue puji gongzuo*, no. 2 (1951): 29.

20. Editorial, "Guangfan chuanbo laodong renmin de shengchan jingyan 广泛传播劳动人民的生产经验 (Disseminating extensively the experiences of laboring people in economic production)," *Kexue puji gongzuo*, no. 2 (1951): 30.

21. Schmalzer, *The People's Peking Man*, 29–31. Between 1972 and 1978 the pictorial assumed different names: *Science Dissemination Materials* (*Kexue puji ziliao*, 1972) and *Science Dissemination* (*Kexue puji*, 1974).

22. Kexue dazhong she 科学大众社, "*Kexue dazhong* bianji gongzuo de chubu ziwo jiantao 《科学大众》编辑工作的初步自我检讨 (A preliminary self-criticism of *Scientific China Monthly* editing team)," *Kexue puji tongxun*, no. 4 (1950): 70.

23. Gu Chaohao 谷超豪, "Banli baozhi kexue fukan de jidian tihui 办理报纸科学副刊的几点体会 (Several points of handling newspaper supplements of popular science)," *Kexue puji tongxun* 1 (1950): 13.

24. Nan Weijun 南玮君, "Shandong dazhong ribao 'kexue zhishi' fukan de bianji gongzuo 山东大众日报'.科学知识'副刊的编辑工作 (The editing work of the *Shandong Popular Daily*'s supplement on scientific knowledge)," *Kexue puji gongzuo*, no. 1 (1951): 14–19.

25. For example, "Beida tongxue hanjia gongzuo de jingyan 北大同学寒假工作的经验 (Beijing University's students' work experience in the winter vacation)," *Kexue puji tongxun*, no. 1 (1950): 11; and "Dakai shenmi xuefu, kexue guihuan renmin! 打开神秘学府，科学归还人民！ (Open the mysterious educational establishment, return science to the people!)," *Kexue puji tongxun*, no. 3 (1950): 40.

26. Hua Luogeng 华罗庚, "Lüetan wo dui shuxue de renshi 略谈我对数学的认识 (On my understanding of mathematics)," and "He tongxuemen tantan xuexi shuxue 和同学们谈谈学习数学 (Discussing the learning of mathematics with students)," *Gei qingnian shuxuejia* 给青年数学家 (To young mathmaticians), ed. Hua Luogeng 华罗庚 (Beijing: Zhongguo qingnian chubanshe, 1956), 11–18. The former was originally published in *Renmin Ribao* on September 21, 1953, the latter in the first issue of *Zhongxuesheng* in 1955.

27. Liu Bangfan 刘邦凡, "Hua Luogeng kepu jiaoyu sixiang yanjiu 华罗庚科普教育思想研究 (A study on Hua Luogeng's educational thoughts on science popularization)," *Qinghai minzu shizhuan xuebao*, no. 1 (1999): 1–5; Hua Luogeng 华罗庚, *Tongchou fangfa pinghua ji buchong* 统筹方法平话及补充 (Popular stories and supplements of CPM methods). (Beijing: Zhongguo gongye chubanshe, 1965).

28. "Wei kaizhan kexue xuanchuan zhong de aiguozhuyi jiaoyu er douzheng 为开展科学宣传中的爱国主义教育而斗争 (Struggle for propagating patriotism in science dissemination)," *Kexue puji gongzuo*, no. 1 (1951): 2–3.

29. Liu Lanshan 刘岚山, "Kan zhanlan qu 看展览去," *Renmin Ribao*, September 18, 1958.

30. Steven D. Lavine and Ivan Karp, "Introduction: Museums and Multiculturalism," in *Exhibiting Cultures: The Poetics and Politics of Museum Display*, ed. Steven D. Lavine and Ivan Karp (Washington: Smithsonian Institution Press, 1991), 1–9.

31. Shida tongxunzu 师大通讯组, "Ji shoudu chunjie kexue zhishi zhanlanhui 记首都春节科学知识展览会 (A report on an scientific knowledge exhibition in the capital during the Spring Festival)," *Kexue puji tongxun*, no. 1 (1950): 4–6; "Shoudu chunjie kezhan bimu qunzhong gongda shiwan yu ren 首都春节科展闭幕群众共达十万余人 (The number of visitors at the end of the Spring Festival Science Exhibition amount to more than 100,000 people)," *Renmin Ribao*, March 1, 1950.

32. "Pingyuan sheng Lin xian wenhuaguan de kexue xuanchuan 平原省林县文化馆的科学宣传 (The science dissemination of the culture center of Lin County in Pingyuan Province)," *Kexue puji tongxun*, no. 8 (1950): 178.

33. Gross, *Farewell to the God of Plague*, 93–94. In the case of using the exhibition to fight against superstition, Denise Ho says that the exhibition's methods of material evidence could hardly present the absent supernatural. See Ho, *Curating Revolution*, 132.

34. "Guizhou sheng kepu xiehui chouweihui yewu gongzuo baogao zhaiyao 贵州省科普协会筹委会业务工作报告摘要 (Excerpt of the report on the work of the Guizhou Province Science Popularization Association Organizing Committee)," *Kexue puji gongzuo*, no. 10 (1951): 208.

35. Chen Hongjie, "Zhongguo jindai kepu jiaoyu: shetuan, changguan he jishu," 29.

36. Wang Lunxin 王伦信, "Minguo shiqi de gonggong kexueguan yu zhongxue like shiyan jiaoxue 民国时期的公共科学馆与中学理科实验教学 (Public science halls in Republican China and the teaching of experiment in middle school science education)," *Journal of South China Normal University* 华南师范大学学报 (社会科学版), no. 10 (2007): 89–94.

37. Chen Hongjie, "Zhongguo jindai kepu jiaoyu: shetuan, changguan he jishu," 38.

38. Xie Junmin 解俊民, "Shenme shi renmin kexueguan 什么是人民科学馆 (What is the People's Science Hall?)," *Kexue puji gongzuo*, no. 1 (1951): 20. Many of the lecture manuscripts were collected in the book series *Dazhong kexue jiangzuo* 大众科学讲座, published by Kaiming Shudian in Shanghai in 1951.

39. Zhongyang renmin kexueguan 中央人民科学馆, *Kang Mei yuan Chao yundong zhong de dongbei yu Chaoxian tuji* 抗美援朝运动中的东北与朝鲜图集 (Atlas of the Northeast and Korea in the Campaign of Resisting America and Aiding Korea) (Shanghai: Zhonghua shuju,1951).

40. Jin Ruicui 金瑞萃, "Beijing shi juban 'Dongbei yu Chaoxian' zhanlanhui 北京市举办"东北与朝鲜"展览会 (The Exhibition of "The Northeast and Korea" Was Held in Beijing)," *Kexue puji gongzuo*, no. 1 (1951): 11–12.

41. Zhongyang renmin kexueguan *Kang Mei yuan Chao yundong zhong de dongbei yu Chaoxian tuji* 抗美援朝运动中的东北与朝鲜图集, 12.

42. Zhongyang renmin kexueguan, *Kang Mei yuan Chao yundong zhong de dongbei yu Chaoxian tuji*, 51; Jin Ruicui 金瑞萃, "Beijing shi juban 'Dongbei yu Chaoxian' zhanlanhui," 12.

43. Bi Yuan 毕苑, *Jianzao changshi: jiaokeshu yu jindai Zhongguo wenhua zhuanxing* 建造常识： 教科书与近代中国文化转型 (Establishing Common Knowledge: Textbooks and the Transformation of Modern Chinese Culture) (Fuzhou: Fujian jiaoyu chubanshe, 2010).

44. Yang Li 杨力, Gao Guangyuan 高广元, and Zhu Jianzhong 朱建中, *Zhongguo kejiao dianying fazhan shi* 中国科教电影发展史 (History of Development of Science Education Films in China) (Shanghai: Fudan University Press, 2010), 7–20.

45. Yang, Gao, and Zhu, *Zhongguo kejiao dianying fazhan shi*, 23–30; Matthew Johnson, "The Science Education Film: Cinematizing Technocracy and Internationalizing Development," *Journal of Chinese Cinemas* special issue, "The Missing Period of PRC Cinema" 5, no. 1 (2011): 36.

46. "Jieshao 'tu dianying' 介绍土电影 (Introducing 'the homemade movie')," *Kexue puji tongxun*, no. 3 (1950): 50; "Huandeng shi kexue puji gongzuo de yige youxiao gongju 幻灯是科学普及工作的一个有效工具 (Lantern slides are an effective tool of science dissemination work)," *Kexue puji tongxun*, no. 9 (1950): 184; "Ji huandeng gongzuo taolunhui 记幻灯工作讨论会 (A report on a symposium on lantern slides work)," *Kexue puji tongxun*, no. 9 (1950): 184–86.

47. The titles were published in a number of lists in *Kexue puji gongzuo*, 1951.

48. Yang, Gao, and Zhu, *Zhongguo kejiao dianying fazhan shi*, 36. In the early 1960s the Society developed into Agriculture Film Studio (*Nongye dianying zhipianchang* 农业电影制片厂) and produced science education films on agriculture, see ibid., 98.

49. "Zhongyang renmin zhengfu zhengwuyuan—Guanyu jiaqiang dianying zhipian gongzuo de jueding 中央人民政府政务院—关于加强电影制片工作的决定 (The directive of the State Council of China's People's Government on strengthening filmmaking work)," *Renmin Ribao*, January 12, 1954.

50. "Zhongguo dianying faxing gongsi jiang juban kexue jiaoyupian zhanlan 中国电影发行公司将举办科学教育片展览 (Chinese film distribution company will hold a science education film exhibition)," *Renmin Ribao*, April 2, 1954. Hao (born 1935) invented the Hao Jianxiu Fine-Yarn Work Method and was named National Industrial Model Worker in 1951.

51. Yang, Gao, and Zhu, *Zhongguo kejiao dianying fazhan shi*, 53–73.

52. Yang, Gao, and Zhu, *Zhongguo kejiao dianying fazhan shi*, 110.

53. Johnson, "The Science Education Film," 42; see also Zhao Huikang 赵惠康, "Xin Zhongguo kejiao dianying de liangci gaochao 新中国科教电影的两次高潮 (Two upsurges of science education movies in the new China)," *Dianying yishu*, no. 6 (2005): 155–60. On the predominance of science education films in the film production during the Mao era see also the findings of Ishii Yumi 石井弓, *Kioku toshite no Nitchū sensō: intabyū ni yoru tasha rikai no kanōsei* 記憶としての日中戦争:

インタビューによる他者理解の可能性 (The Japanese-Chinese War as memory: possibility of understanding others by interview) (Tōkyō: Kenbun shuppan, 2013).

54. Situ Huimin 司徒慧敏, "Changkai kexue zhishi de damen—zhu guochan kexue jiaoyu yingpian zhanlan chenggong 敞开科学知识的大门—祝国产科学教育影片展览成功 (Widely opening the doors of scientific knowledge—we wish the exhibition of Chinese science education film much success)," *Renmin Ribao*, June 7, 1963.

55. Yang, Gao, and Zhu, *Zhongguo kejiao dianying fazhan shi*, 81.

56. Jilupian zhi chuang: Heibai kejiaopian "Jin Xiaofeng yu hong lingchong" 纪录片之窗,"黑白科教片《金小蜂与红铃虫》" (Black and white science education film *The Jewel Wasp and the Pink Bollworm*) (1963), YouTube, accessed November 15, 2014, https://www.youtube.com/watch?v=lRecIndANxY.

57. Schmalzer, *Red Revolution, Green Revolution*, 38 and 42, 109.

58. Situ Huimin, "Changkai kexue zhishi de damen."

59. Johnson, "The Science Education Film," 47. See also Fan Ka-wai, "Film Propaganda and the Anti-schistosomiasis Campaign in Communist China," *Sungkyun Journal of East Asian Studies* 12, no. 1 (April 2012): 1–17.

60. Xu Xiaxiang 徐霞翔, "Toushi nongcun dianying fangyingyuan—yi ershi shiji wushi niandai Jiangsu sheng wei li 透视农村电影放映员—以二十世纪五十年代江苏省为例 (The film project team in rural areas—taking Jiangsu Province of the 1950s as an example)," *The Twentieth Century* (Web version), published March 31, 2009, last accessed December 20, 2020, http://www.cuhk.edu.hk/ics/21c/media/online/0812018.pdf.

61. Guangdong Xinhui xian dianying guanlizhan 广东新会县电影管理站, "Ba kexue zhishi chuanbo gei guangda nongmin 把科学知识传播给广大农民 (Disseminating scientific knowledge amongst peasants)," *Dianying yishu*, no. 6 (1965): 78–80.

62. "Nongcun dianying yingqian xuanchuan gongzuo shishi banfa 农村电影映前宣传工作实施办法 (Methods of propaganda work before projecting movies in rural areas)," *Dianying puji*, no. 6 (1982): 15–16.

63. "Fanghao kejiao dianying, cujin shengchan fazhan 放好科教电影，促进生产发展 (Screening science education movies to increase productivity)," *Dianying puji*, no. 5 (1982): 4–7.

64. Zhou Wenxiang 周文樁, "Qin'ai de pengyoumen! 亲爱的朋友们 (Dear friends!)," *Zhishi jiushi liliang*, no. 1 (1956): 2.

65. Cf. Sonja Schmid, "Celebrating Tomorrow Today: The Peaceful Atom on Display in the Soviet Union," *Social Studies of Science* 36, no. 3 (2006): 356.

66. Ian Welsh, *Mobilising Modernity: The Nuclear Moment* (London & New York: Routledge, 2000): 18, quoted by Schmid, "Celebrating Tomorrow Today," 356. Italics added.

67. Brigitte Schroeder-Gudehus and David Cloutier, "Popularizing Science and Technology During the Cold War: Brussels 1958," in *Fair Representation: World's Fairs and the Modern World*, eds. Robert W. Rydell and Nancy Gwinn (Amsterdam: VU University Press, 1994), 157–80.

68. *Knowledge Is Power* (*Zhishi jiushi liliang*), no. 9 (1957).

69. A. Agarkow, E. Leonenko, and G. Müller, "Shaft Drilling in the U.S.S.R.: History and Recent Experiences," in *Shaft Engineering*, ed. Institute of Mining and Metallurgy (London: Taylor & Francis e-Library, 2005), 1–32. The print version of the book came out in 1989.

70. See for example, "Xinxiang 信箱," *Kexue huabao* 1, no. 3 (1933): 117.

71. "Wenti jianda 问题简答," *Kexue huabao* 21, no. 7 (1953), back cover.

72. For example, a forensic explanation of "zombie" (*jiangshi* 僵尸) was offered in *Kexue huabao*, no. 12 (1957): 466.

73. For example, the story about Huang Daopo 黄道婆, the woman in the thirteenth century who improved the weaving tools and was said to have spread her weaving skills among her contemporaries, *Kexue huabao*, no. 6 (1957): 236.

74. For example, the report and explanation of solar halos, *Kexue huabao*, no. 5 (1957): 195.

75. "Dajia tan 大家谈 ," *Kexue huabao*, no. 7 (1957): 316–17.

76. Jing Han 静寒, "Naiyang ye hui sheng jiehebing 奶羊也会生结核病 (Goat can also get tuberculosis)," *Kexue huabao*, no. 1 (1963): 37.

Chapter Three

Promising a Bright Future
The (Half-)Mechanization of Agricultural Production

When the Communist Party started to transform China's countryside in the second half of the 1940s, its declared aim was to create the new (wo)man while "liberating" the peasant from the hardships of exploitation. According to the *Pictorial on Land Reform* (*Tudi gaige huaji* 土地改革画集, 1952), the situation of the peasants before 1949 had been miserable and oppressive, with many people either starving or being killed by the landowners. Under the old regime the rural population had been divided in two distinct groups, "landowners living in paradise" and "peasants scorched in hell,"[1] with the latter group having no access to means of production and suffering heavy taxes, fees, and rents to an unprecedented scale. The *Pictorial* declared in bold letters that it was the intention of the CCP to end this inhumane situation by redistributing land among the landless peasants.[2] It visualized the struggle against the landowners of the rural area near Shanghai with horrendous pictures that depict the bloody ruthlessness of the class enemy.[3] It saw expropriation as part of the anti-feudal and anti-capitalist struggle to return the means of production to the peasants and thereby promote the economic development of the country.[4]

The land reform was part of a series of far-reaching socio-political measures to increase productivity. Yet, simply returning the land to the tiller soon proved to be insufficient. The rapidly growing population and the lack of machinery and chemical fertilizers made the yield increase per capita difficult.[5] The CCP thus combined class struggle—the exposure of sabotage and concealed property of rich peasants—with the introduction of new management practices and agricultural technology, ranging from new breeding techniques of crops and animal husbandry to new machinery. The early campaigns of mechanizing agricultural production aimed primarily at relieving monotonous backbreaking toil such as the "three bending" in carrying and pumping, in transplanting seedlings and in harvesting crop. The improvement of the means of production on a national scale was, however, hindered by the lack of resources and capital, despite the Soviet Union's offer to help increasing tractor production in China. In this chapter, we focus on the social and political conditions of knowledge production to show how the idea of experiment and the identification of typical cases (*dianxing*) were used as viable ways to incorporate the creativity of the masses into the overall transformation of the agricultural sector. Whether and to what degree their innovations truly contributed to raising yields is of less interest here. Rather, our concerns are to show that trends of modernization and innovation in the agricultural sector do not necessarily follow the binary logic, according to which radical and technocratic periods alternated, as argued by Schmalzer.[6] In this and the following chapters we also intend to interrogate the binary of *tu* 土 (native, Chinese, local) and *yang* 洋 (foreign, professional, ivory tower) knowledges as an epistemic principle of scientific innovation, whose implications can be felt beyond the revolutionary era.

During the 1950s, the modernization of agricultural production pursued two major aims, namely expanding the acreage under cultivation and raising the yields per *mu*. In the First Five-Year Plan for Development of the National Economy special attention was given to the latter, which was supposed to be achieved by the introduction of new technologies.[7] In January 1956, the CCP adopted an official decision encompassing twelve concrete measures. These were:

1. building up irrigation,
2. increasing the application of fertilizer,
3. improving old-style tools and developing new tools,
4. promoting superior seed strains,
5. expanding multiple-cropping areas,
6. producing more high-yield crops,
7. practicing intensive cultivation and improving farming methods,
8. improving soil, conserving water and soil,
9. intensifying animal husbandry,
10. eliminating pests and plant diseases,
11. reclaiming wasteland, and
12. increasing farming land.

Two years later the Eight-Point Charter of Agriculture (*Nongye bazi xianfa* 农业八字宪法) shortened the twelve measures into eight single, better memorizable characters: landscaping (*tu* 土), fertilizer (*fei* 肥), water conservancy (*shui* 水), seed-strain improvement (*zhong* 种), close planting (*mi* 密), crop protection (*bao* 保), field management (*guan* 管), and tool reform (*gong* 工).[8]

These measures intended to transform agricultural production on a large scale. While science dissemination materials in the early to mid-1950s emphasized the model character of Soviet agriculture and propagated the tractor as the icon of agricultural mechanization, technological reality in Chinese villages looked differently, as this chapter is going to show. Focusing on the creation and dissemination of agricultural knowledge, especially the changes and conflicts in the introduction of Soviet agricultural technologies in the first years of the PRC, and on the development of *tu*, or indigenous, technologies during the Great Leap Forward that valorized local experiments as empirical innovation at the grassroots level, we show that tool improvements were a central element in agricultural mechanization. While the tractor clearly embodied a dream of mechanization as the *yang* 洋 ideal, the examples presented in this chapter—the case of the half-mechanized two-wheel, two-blade (TWTB) plow and the local appropriation of wheelbarrows and ball-bearings by *tu* 土 methods—show that the process of introducing new technology was more complex than importing or manufacturing agricultural machines in a copy-paste fashion. The analysis of that complexity allows insights into the process of knowledge production where more than one actor was present.

THE SOVIET TRACTOR AS THE ICON OF FULL AGRICULTURAL MECHANIZATION

In its early years, or more precisely, starting in 1950, the People's Republic oriented itself to the Soviet model, be it by sending chosen citizens to the Soviet Union to receive training in various fields of industry and agriculture, or by welcoming Soviet advisors to China. The exact numbers of both groups are difficult to determine, ranging from several hundreds to ten thousand from 1950 to 1960.[9] In case of Soviet advisors, Kaple holds that the actual exchange started after Stalin's death in 1953, while the first group of Chinese students was sent to the Soviet Union in 1951, with an estimated number of 16,000 students before the Sino-Soviet split in 1961.[10]

Before 1953, knowledge transfer took place primarily in the form of translated books and journals. From late 1949 on there was a bustling translation of Soviet books and manuals on science and technology.[11] Chinese communists eagerly read Soviet newspapers, journals, and books to obtain valuable insights in socialist construction. The *Chinese General Title Catalogue* (*Quanguo zongshumu* 全国总书目) for the years of 1949 to 1953 includes thousands of Soviet books that were translated and printed in Chinese.[12]

Before Mao Zedong issued his warning against "mechanical absorption of foreign material" in his speech "On the Ten Major Relationships" delivered in April 1956, the Party took over in the almost wholesale fashion the Soviet model of economic development in the First Five-Year Plan (1953–57), as shows an internal reading material (*neibu duwu* 内部读物) published one month later. It includes reports of the divisions of the Chinese Academy of Sciences as well as five editorial pieces taken from the *People's Daily* (1954–55) that unanimously accept the superiority of the Soviet Union in scientific research.[13] In a similar fashion a booklet titled *Learning from the Soviet Experts* (*Xiang Sulian zhuanjia xuexi* 向苏联专家学习, 1953) expresses in every chapter gratitude for the Soviet help to speed up China's industrialization, in all sectors ranging from steel and petroleum to papermaking and veterinary medicine.[14]

Among the numerous publications introducing advanced agricultural knowledge to the Chinese audience, the most exemplary one is the journal *Soviet Agricultural Science* (*Sulian nongye kexue* 苏联农业科学, 1950–58). It provided translations of Soviet articles dealing with new discoveries in plant breeding, new methods in herbicide and pesticide use as well as breakthroughs in animal husbandry. Later publications—either direct translations or summaries of various Soviet publications—introduced mechanization and agricultural machines.[15] Mechanization was understood here as increasing labor efficiency through automation of work processes, improvement of the motorized drive, and the construction of machines that could perform multiple tasks at once.

The most iconic machine that promised saving labor power was the tractor. First imported from the Soviet Union and other countries of the Warsaw Pact, they were believed to provide the most effective means to increase productivity.[16] A look at the introductory chapters to manuals and tutorials dealing with tractors and other machines from the 1950s shows that mechanization was viewed as a necessary step to achieve modernization.[17] Following historical materialism it is seen as the most suitable path for redeeming China from backwardness and poverty after exploitative imperialism and KMT's reactionary rule.

In 1954, the All-China Association for the Dissemination of Scientific and Technological Knowledge published a book titled *Tractors and Combine Harvesters* (*Tuolaji he lianhe shougeji* 拖拉机和联合收割机). Written by Xu Guohua and Liu Jiansheng, it introduced combine harvesters—or *kangbaiyin* harvesters (康拜因, similar to the English word to combine, or the Russian term комбайн)—as the new breakthrough in mechanization. It integrated seven separate steps involved in harvesting, threshing, and collecting grain into a single machine, thereby increasing both harvesting speed and productivity.[18] As an effect thereof, peasants could enjoy a higher life quality, the authors argued. This logic is most succinctly illustrated by a graph in the book, where the use of various technologies for plowing, harrowing, sowing, and harvesting are compared in order to convince the reader of the superiority of tractors: while a single peasant can dig up 0.2 *mu* in ten hours using a manual hoe, a DT-54 Soviet-made tractor (54 hp engine, built from 1949 to 1979) with its five blades can plow 100 *mu* in the same amount of time, representing an increase by the factor 500. Similar are the numerical increases in harrowing, sowing, and harvesting.[19]

The first Chinese tractor factory was planned in July 1953, yet it did not start mass production until 1959. The planned number was 15,000 units in that year, which was far from sufficient when taking into account the size of the arable land of the whole country. Though the First Five-Year Plan entailed measures to make more machinery available, the simultaneous prioritization of industrial development turned out to be an obstacle to the achievement of mechanization on a larger scale. Mao Zedong commented during a conference of secretaries of provincial, municipal, and autonomous region party committees on July 31, 1955, that

> some of our comrades have also failed to give any thought to the connection between two other facts, namely, that large funds are needed to accomplish both national industrialization and the technical transformation of agriculture, and that a considerable part of these funds has to be accumulated through agriculture.[20]

Thus, the sector of agriculture not only suffered the general lack of funds but also was squeezed to fund industrialization. As a consequence, even the import of tractors became difficult (if the acquisition of a large amount of machinery had been possible at all).[21] The capital drain in favor of industrialization caused Mao to predict that a widespread technical transformation of the agricultural sector would at least take twenty-five years and indeed through collectivization—only then were peasants able to accumulate sufficient capital and labor to mechanize: "In agriculture, with conditions as they are in our country co-operation must precede the use of big machinery."[22]

Despite the financial constraints and despite the country's topography of the arable land that made a nation-wide tractor use difficult (while the north-east and the north of the country had suitable terrains, the south with its small rice paddies and hilly areas had not), Soviet experts convinced Mao and his cadres that big tractors were the true icons of socialist agriculture. This led to the adoption of a ten-year mechanization program in 1958 and the establishment of the Ministry of Agricultural Machinery (*Nongye jixiebu* 农业机械部) in August 1959. Under the leadership of Chen Zhengren

陈正人 (1907–72), the Ministry met Mao's demand to open several hundred farm machinery research bureaus (*nongju yanjiusuo* 农具研究所) at the prefectural and county level.[23]

The success of these measures was less than expected, because Liu Shaoqi 刘少奇 (1898–1969) and Mao could not agree on the ownership and management of machines. While Mao preferred collectives, Liu considered state-controlled tractor stations to be more promising: the collectives had neither the financial means to buy tractors nor sufficient technical knowledge to maintain and operate the machinery. Their struggle followed the well-known lines of mass voluntarism vs. technocratic expertise. After Mao had received reports on the Soviet agricultural mechanization, the party voted at a national conference in 1958 for the transfer of tractors to the people's communes.[24] As a result, these were no longer under the supervision of technical experts. Technical breakdowns and the assignment of maintenance to the lay(wo)man reduced the efficiency of machinery.

These problems were seen, however, not primarily as a result of technological backwardness and economic constraints, but rather as a problem of politics. In the propaganda of the 1950s, the promotion of local industry to provide farming tools was often combined with nurturing political consciousness. For instance, in 1953 a certain Leng Quan 冷泉 argued that the socialist emulation campaign was successful in lowering agricultural production costs and fighting against waste.[25] A farm in Xinyang in Henan Province reported in the same year that four technicians succeeded in repairing a combine harvester that had been broken for three years. In addition, they managed to spend considerably less for the repair than estimated (a mere 20,000 Yuan compared to the estimated 10 million Yuan).[26]

A few years later a book titled *How to Increase the Working Efficiency of Tractors* (*Zenyang tigao tuolaji de gongzuo xiaolü* 怎样提高拖拉机的工作效率) lauded the widespread use of tractors in agricultural cooperatives. Addressing workers on farms and tractor stations, the book emphasized the need to increase work efficiency while reducing accidents and machine damages. In order to do so, it states, modern and progressive tractor stations should be established throughout the country. The book chapters were written in the style of personal accounts, presenting the personal experiences of workers of the farms and tractor stations as typical cases that could be imitated. In addition to the widespread emphasis on the correct political consciousness and commitment to the production process, a number of hints addressed organizational and technical aspects. For instance, the Wutonghe farm in Heilongjiang Province stressed that labor division and a functioning accountability system were imperative. Cleaning and lubrication of the engine should be done every thirty—instead of the usual sixty—hours of operation, and special attention should be paid to the abrasion of moving parts and the axis state. In addition, oil pressure and engine noise must be checked at each shift. Only then could premature wear or accidents be avoided.[27] Such a cautious attitude toward machinery can be explained either by the constant call for increasing production with less cost (*zengchan jieyue yundong* 增产节约运动), a central slogan in the 1950s and 1960s, or by the very fact that financial and materiel constraints prevented an immediate overall availability of advanced machinery as imagined in the Soviet Union. In the end, both factors led to a different approach in mechanizing agriculture, in which the decision to implement mechanical power was delegated to local actors. However, this did not prevent political intervention from occurring and in some cases proved to be counterproductive.

THE TWO-WHEEL TWO-BLADE PLOW AS AN IMPORTED TOOL OF MECHANIZATION

In comparison with the tractor, the introduction of the two-wheel, two-blade (TWTB) plow (*shuanglun shuanghuali* 双轮双铧犁) from the Soviet Union was more contested. Imported during the First Five-Year Plan, the nationwide promotion of the TWTB plow was considered to be one key step before realizing full agricultural mechanization, as declared in an editorial of *People's Daily* on January 6, 1955:

> To realize socialist transformation in agriculture in our country, we need two steps: first we must realize cooperative; and secondly, we realize mechanization. And the realization of the mechanization also requires two steps: before we are able to produce large quantities of tractors and gasoline, we can only forcefully promote new animal-pulled agricultural tools such as TWTB plow and two-wheel, single-blade plow on the basis of installing cooperatives. First, we improve the agricultural tools and then, when we realize all necessary conditions, can we realize large-scale mechanization.[28]

To simply take over a foreign plow was not easy because it was more complex machinery. Designed for cultivating sandy or light clay soils in a faster and deeper manner, it suited neither China's terraced and small farm plots, such as the rice fields of South China, nor plots with heavy clay soil. Moreover, the tool was reportedly too complex to be used by peasants with limited technical knowledge and far too expensive for individual households.[29] Despite these disadvantages the TWTB plow was promoted with full force, with Mao Zedong dictating in the famous *Forty Articles of Agriculture* (*Nongye sishi tiao* 农业四十条) of 1956 that six million TWTB plows should be built and distributed within three to five years.³⁰ This plan was opposed by Zhou Enlai and Liu Shaoqi who warned against rash advance (*maojin* 冒进), criticizing the spread of the TWTB plow in South China as a consequence of unrealistic economic planning.³¹ As the result the plows, of which 1.7 million had been produced in 1956 alone, ended up—instead of being used—hanging on the wall in famers' houses (*gua zai qiangshang wufa shiyong de "guali"* 挂在墙上无法使用的"挂犁").³² The situation changed in 1957 when Mao resumed his power during the Anti-Rightist Campaign, forcing Zhou Enlai to step back and allow the Great Leap to take off.

Political pressure from the top notwithstanding, the TWTB met resistance in the countryside where the so-called latest technological achievement went against proven farming practices. In case of peasants in Liaoning Province Alfred Chan shows that the Party's 1958 campaign for deep plowing in fact caused its alienation from the peasantry. By promoting deep plowing as a way to raise crop yield by up to five times, the Party's demand brought about additional hardship (also due to the lack of tractors and other machinery), and in some cases even resulted in moving infertile sub-soil to the top. For the Liaoning Provincial Party Committee, however, the commitment to plowing meant to prove the sincerity of one's revolutionary consciousness.³³ On April 16, 1958, the *People's Daily* attributed the failure of disseminating the TWTB plow to "reactionary undercurrent":

> While the system of means of production was undergoing a transition from private ownership to collective ownership, and the two-wheel two-blade plow was spread to replace their old-fashioned counterparts and consolidate the ownership of the collectives, the resentment toward socialism among certain groups of rich peasants, the conservative thought and inertia of peasants, [. . .] together with the storms and waves stirred by rightists, had merged into a reactionary undercurrent [. . .]. This was the main reason for which the two-wheel two-blade plow could not be widely employed.³⁴

A technical problem was interpreted as a struggle between two political lines that resulted in the summer of 1958 in the call for "restoring the reputation of TWTB plow" (*wei shuanglun shuanghuali huifu mingyu* 为双轮双铧犁恢复名誉).³⁵ The campaign ended in late 1958, as Chan shows, because the grassroots units were not able to manufacture large numbers of plows, including a cable-drawn variant, by using their own resources.³⁶

As part of the Great Leap Forward, the deep-plowing campaign is one striking example of how ideology penetrated the daily life of peasants. However, it would be too simple to reduce it to political pressure. Instead, we propose to go beyond the idea of an all-pervading two-line struggle in order to focus on how different forms of knowledge were justified by references to science. Leaving aside the exaggerated demands of "the deeper the better," the TWTB was promoted for its assumed superior scientific and/or modern quality. Though the state's ignorance of proven local practices complicated the acceptance of the TWTB, this did not mean that the peasants were always reluctant to change their production techniques. This was achieved by the identification of typical cases (*dianxing*), which valorized local knowledge and the possibility of its being spread and adapted as a viable way to incorporate the creativity of the masses.

GOING BACK TO THE GRASSROOTS: LOCALIZING BALL-BEARINGS AND WHEELBARROWS

The ideal of nation-wide mechanization in agriculture faced considerable obstacles due to limited resources and regional differences. This pushed the Central Politbureau at the Chengdu Conference in March 1958 to promote a new model called technological reform (*jishu gaizao* 技术改造) that prioritized intermediate technology. On March 15, 1958, the Ministry of Agriculture, the First Ministry of Commerce, the General Office of the People's Bank, and the All-China Federation of Handicraft Producers' Cooperatives jointly issued a directive that entailed two approaches. It demanded that new, multi-purpose tools should be designed, produced, and funded locally while at the same time calling for the improvement of existing tools. In addition, mechanization should make use of locally available energy sources such as wind and water power, next to manpower and electricity.[37] A few months later—after the directive failed to generate the expected enthusiasm among the farmers—Peng Zhen 彭真 (1902–97), then mayor of Beijing, and Tan Zhenlin 谭震林 (1902–83), the member of the 8th Politburo of the CCP and the responsible cadre for agriculture during the Great Leap Forward, telephoned provincial Party committees on August 20, 1958. Instructed by Mao, they admonished the lack of revolutionary spirit.[38]

One technological improvement that gained particular attention in that summer were ball-bearings. Two days prior to the aforementioned telephone calls the *People's Daily* had already stressed the urgent need to start their construction, no matter what the conditions were. Comparing two neighboring counties in Shaanxi Province—Xianyang county 咸阳县 and Xingping county 兴平县—the article pointed out that although both counties had farming tools construction bureaus Xianyang county relied on the masses and therefore had achieved far greater success in procuring the intermediate technology.[39]

Ball-bearings (*gunzhu zhoucheng* 滚珠轴承) were intended to contribute to mechanization by decreasing rotational friction while supporting axial loads. They should be integrated into already existing transport tools such as carriages and wheelbarrows, or processing machines and waterwheels, making them more efficient and robust, as a special issue of *Popular Science* (*Kexue dazhong*) presented in September 1958, which discussed ball-bearings in numerous articles and propagated them as key to half-mechanization (*ban jiexiehua* 半机械化). In a 1959 party communiqué on tool reform Mao Zedong described such mechanization to be in principle initiated by experts, but ultimately carried out and propelled forward by the masses—instead of relying on tools and machines provided by the state or delivered by foreign friends.[40]

Booklets such as *Tools for Plowing and Weeding Created by the Masses* (*Qunzhong chuangzao de zhonggeng chucao nongju* 群众创造的中耕除草农具, 1958) depict the Chinese farmers as more creative and more industrious than their counterparts in capitalist states.[41] Other publications list technological innovations collected from counties and provinces all over the country. The twenty-volume manual *Selected Blueprints of Farming Tools* (*Nongju tuxuan* 农具图选) of 1958 gathered construction plans of tools, including information on the origin and distribution of the tool, i.e., naming the county and in some cases the farmer who invented it. The innovations were then disseminated so that typical cases (*dianxing*) of local origin could serve a national cause.[42] The focus of the blueprints were tools for irrigation, seedling transplantation, and transportation (including those for moving of earth for the campaigns of increasing arable land and constructing water dams), that is primarily the backbreaking toil of the "three bending."

In fact the general shortage of steel, coal, and gasoline during the Great Leap Forward years demanded technological self-reliance, which was the primary reason why emphasis was rather put on the improvement of already existing farming tools.[43] The simultaneous emphasis on the de-professionalization of mechanization led to the establishment of local farm machinery research bureaus in each county of each province, where technicians, experienced carpenters, and blacksmiths should join forces with the farmers to create and/or improve tools.[44] In the case of ball-bearings, this led to a frenzy in devising creative alternatives to the technological standard that had been set forth by urban experts and Soviet advisors. Anna Louise Strong reports in her observations of the people's communes that metal bearings contained balls made of glass, porcelain, or even acorns (extraordinarily round acorns were a specialty of Sichuanese oaks). For acorn bearings, the races were made of bamboo. The need to devise non-standardized ball-bearings—especially with regard to their size—resulted from the very fact, Strong revealed, that existing wheelbarrows, carts, and waterwheels came in an unlimited variety; industrially produced, standardized bearings would simply not fit.[45]

The government started to invest heavily in propagating half-mechanization by emphasizing its contribution to raise productivity. In 1958 a copy of science dissemination material called *Using Local Methods to Achieve Half-Mechanization* (*Yong tu banfa shixian banjixiehua* 用土办法实现半机械化) pointed to the significant growth of transport capacity when advancing wheelbarrows by locally produced ball-bearings. While a farmer could push a weight of 200 *jin* (approx. 220 pounds) in a conventional wheelbarrow, an advanced wheelbarrow with ball-bearings increased the capacity to 600 *jin*.[46] The booklet insisted that such an innovation could be developed with limited local resources by pointing to the achievements of Suqian County, where twenty-three ball-bearing factories had reportedly been established that constructed the necessary tools and manufactured 30,000 ball-bearings on a daily basis. According to the authors, the success clearly refuted the superstitious belief that ball-bearings could only be imported.[47]

The same assessment with regard to the rise of productivity is made in a 1958 catalogue of farming tools that had been compiled on the occasion of a nation-wide exhibition in 1958.[48] It lists a large variety of farming tools for irrigation, plowing, transportation, etc. They are introduced as improvements of previously existing tools, with each construction plan naming the inventor and his work unit. The purpose of the tool and its construction are described in detail, including information on the materials and estimated costs.[49] The most striking example in the efforts to adapt tools to local circumstances is a heavy-duty wheelbarrow made in five different size versions, with the biggest one suitable for adult males transporting 800 to 1,000 *jin* of soil, and the smaller one for children between the ages of seven and ten carrying 120 *jin*.[50]

The construction of wheelbarrows for children itself is not an exceptional phenomenon. Participation of children in labor was—and today still is—quite common among peasants in non-industrialized countries.[51] The efforts to provide tools for physically less strong producers also included the "Women's Leap Forward handcart" (*Funü yuejinche* 妇女跃进车) that reportedly enjoyed great popularity in the countryside, as the catalogue boasted.[52]

The 1959 propaganda film *A Handcart with Steel Balls* (*Gangzhu feiche* 钢珠飞车) combined the celebration of the innovative approach of improving tools with local means with the question of emancipation. Describing a competition in transporting grain to the state granary during the Great Leap Forward, it details the ambition of the female workforce to surpass the men of their village despite having less physical power. At the end of the film they win in the men vs. women competition by having the technologically superior wheelbarrows with integrated ball-bearings. Their success is explained by

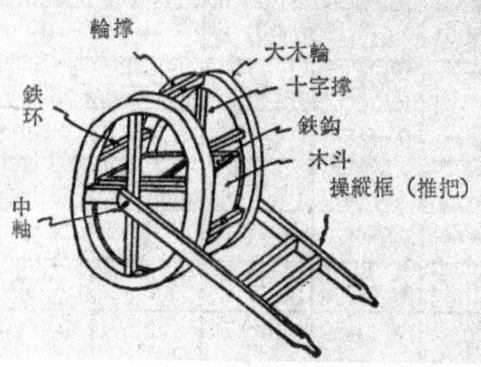

Figure 3.1. Quanguo nongju zhanlanhui 全国农具展览会, ed. *Quanguo nongju zhanlanhui—tuijian zhanpin—nongtian paiguan jixie* 全国农具展览会—推荐展品—农田排灌机械 (National Exhibition of Agricultural Tools—Recommended Exhibits—Farmland Irrigation Machines). Beijing: Kexue puji chubanshe, 1958.

their perseverance in experimenting with different ways of forging the necessary steel balls.[53]

The ideological empowerment of the masses in the creation and dissemination of science and technology was in the majority of construction plans legitimized by making "experience" (*jingyan*) an epistemological category. The journal *Agricultural Science of Guangxi Province* reported that with the establishment of the people's communes a new emphasis on high-yield and experimental plots created a large group of scientific personnel (*kexue*

附表2　　　　　　　　　　**重力滾車載重表**

品　種	可裝物料重量				備　注
	土(斤)	砂(斤)	石(斤)	糞(斤)	
一号車	800—1,000	800—1,000	800—1,000	700	適用于一般整勞動力
二号車	500—600	500—600	500—600	300—400	適用一般勞力（即15—17岁男、女，和老年人)
三号車	200—250	200—250	200—250	200以上	適用于半勞力和13—15岁男女
四号車	約200	約200	約200	150	適用10—12岁兒童和妇女
五号車	120	120	120	80—100	適用于7岁以上兒童

附表3　　　　　　　　　　**重力滾車价格表**

品　種	所需材料、人工						工数	金額	总計	說　明
	木　板		鉄　釘		鉄　絲					
	数量	金額	数量	金額	数量	金額				
一号車	30尺	8.7	1.4斤	1.46	1.8斤	0.9	8	8.00	19.06	
二号車	23尺	6.6	1.1斤	1.1	1斤	0.6	7	7.00	15.3	
三号車	30尺	6	1斤	0.78	1斤	0.6	6	6.00	13.38	
四号車	17尺	5.1	1斤	0.78	1斤	0.6	5	5.00	11.45	
五号車	15尺	4.5	0.12斤	0.45	0.12斤	0.45	4	4.00	9.4	

注：①木板的价格是以混合价計算，按自产自銷不加稅額；
　　②木工每个按1元計算；
　　③如用社內木料，木工按工分計算，仅鉄料开支以二号車算，仅用2元多錢就可以了。

Figure 3.2. Quanguo nongju zhanlanhui 全国农具展览会, ed. *Quanguo nongju zhanlanhui—tuijian zhanpin—nongtian paiguan jixie* 全国农具展览会—推荐展品—农田排灌机械 (National Exhibition of Agricultural Tools—Recommended Exhibits—Farmland Irrigation Machines). Beijing: Kexue puji chubanshe, 1958.

yanjiu renyuan 科学研究人员), among which local experts (*tu zhuanjia* 土专家) and representatives of the masses achieved good results and proved the superiority of the communes.[54] The definition of advanced science (that is, applied science) depended on its relation to "practice" (*shijian*). Because science had to serve the people, it had to be empirical and utilitarian, and this made it possible to consider indigenous knowledge and practices that did not necessarily conform to the standard notion of modern agricultural sciences.

Though the major cause was economic constraints, the local production of ball-bearings is a classic example of how the government wanted to incorporate scientific knowledge from below.[55] It did so even if it meant defending unconventional wisdom, which is also achieved by avoiding the discussion of rationality or objectivity as epistemological principles in the blueprints and manuals detailing mechanization.

Rather, typical cases are identified as a result of practical experiences. Being deduced from local practices makes generalizations difficult, which again is at odds with the idea of universal science. Yet, it allowed for a larger degree of creativity and more room to accommodate local circumstances. This nevertheless did not reduce the role of the state as the dominant actor in claiming and controlling indigenous knowledge, as argued by Fan Fa-ti:

> The state faced the question of how to manage a body of knowledge that was at the same time problematic (e.g., when it took the form of superstition) and valuable (both scientifically and politically). In collecting, scientizing, and disseminating folk knowledge, the state attempted to control the knowledge for practical as well as political purposes and to transform traditional folk practice into something radically modern in its political meaning and scientific application. In doing this, the state de facto claimed the ownership of a body of knowledge that had previously been in possession of the common folk. It tried to discipline this knowledge by exercising its political authority.[55]

When discussing the contradictory images of the peasant as both experienced and backward, Sigrid Schmalzer confirms Fan's point by pointing out that "the celebration of 'old peasant' as experienced and knowledgeable operated in very narrow constraints."[56] The attempts of control notwithstanding, the valorization of local knowledges, though not conforming to Mao's earlier ideal of following the Soviet model, was also not in line with the more technocratic approach envisioned by Zhou Enlai and Liu Shaoqi. In other words, the introduction of new agricultural technologies emerged as a distinct approach that cannot be explained by reference to the two-line struggle in party leadership exclusively.

To conclude, the various strategies of realizing (half-)mechanization during the Great Leap Forward years—and especially tool improvement—entailed the acknowledgment of local indigenous knowledge based on empirical experiences. It was mobilized because technological development does not always follow immutable laws of science, but depends on local circumstances. Though unquestionably shaped by political decisions, the true historical significance of the turn to the local perhaps lies in the attempt to weaken the binaries of traditional/modern or foreign/indigenous knowledge. In the steel and iron industry—as the following chapter is going to argue—the situation was slightly different because local circumstances were far less important in newly established factories than effective organization and management. To accelerate national industrialization, China searched for ways of combining expert knowledge and knowledge gained by workers on the shopfloor.

NOTES

1. Huadong junzheng weiyuanhui tudi gaige weiyuanhui 华东军政委员会土地改革委员会, *Shanghai shijiaoqu Sunan xingzhengqu - tudi gaige huaji* 上海市郊区苏南行政区—土地改革画集 (Huadong junzheng weiyuanhui tudi gaige weiyuanhui, 1952). Influential in formulating the land reform was Mao Zedong's 1925 *Analysis of the Classes in Chinese Society* (*Zhongguo shehui ge jieji de fenxi* 中国社会各阶级的分析), according to which the five classes were landlords, rich peasants, middle peasants, poor peasants, and farm workers. See Mao Zedong, "Analysis of the Classes in Chinese Society," in *Selected Works of Mao Tse-tung*, Vol. I (Peking: Foreign Languages Press, 1977), 13–21.

2. On the land reform and the expropriation of the landlords see William Hinton, *Shenfan* (New York: Random House, 1983); William Hinton, "The Importance of Land Reform in the Reconstruction of China," *Monthly Review* 50, no. 3 (1998): 147–60; Jan Myrdal, *Bericht aus einem chinesischen Dorf* (München: dtv, 1969); Julia Strauss, "Rethinking Land Reform and Regime Consolidation in the People's Republic of China: The Case of Jiangnan (1950–1952)," in *Rethinking China in the 1950s*, ed. Mechthild Leutner (Berlin: Lit Verlag, 2007), 24–34; Brian DeMare, "Casting (Off) Their Stinking Airs: Chinese Intellectuals and Land Reform, 1946–52," *The China Journal* no. 67 (January 2012): 109–29.

3. Similar to the later propaganda material on the landlord Liu Wencai 刘文彩 (1887–1949). See Denise Ho and Li Jie, "From Landlord Manor to Red Memorabilia: Reincarnations of a Chinese Museum Town," *Modern China* 42, no. 1 (2016): 3–37.

4. The first part of the *Pictorial on Land Reform* deals with the situation on the countryside before the land reform, providing detailed information on land tenure and the social structure. It states that poor peasants make more than 50 percent of the population in the Sunan area, but have only 20 percent of the land while the landowners possess more than one third of the soil, but occupy only 3.25 percent of the population.

5. Judith Shapiro, *Mao's War against Nature—Politics and the Environment in Revolutionary China* (Cambridge: Cambridge University Press, 2001).

6. Schmalzer, *Red Revolution, Green Revolution*, 27–28.

7. See *First Five-Year Plan for Development of the National Economy of the People's Republic of China in 1953–57* (Beijing: Foreign Languages University Press, 1956).

8. For this Charter, see Xu Liangying and Fan Dainian, *Science and Socialist Construction in China* (Armonk: M. E. Sharpe, 1982), and Schmalzer, *Red Revolution, Green Revolution*, 29–34.

9. Leo A. Orleans, "Soviet Influence on China's Higher Education," in *China's Education and the Industrialized World: Studies in Cultural Transfer*, ed. Ruth Hayhoe and Marianne Bastid (Armonk: M. E. Sharpe, 1987), 188; Izabella Goikhman, "Soviet-Chinese Academic Interactions in the 1950s: Questioning the 'Impact-Response' Approach," in *China Learns from the Soviet Union, 1949–Present*, eds. Thomas P. Bernstein and Hua-yu Li (New York: Rowman & Littlefield, 2010), 282; Deborah A. Kaple, "Soviet Advisors in China in the 1950s," in *Brothers in Arms:*

The Rise and Fall of the Sino-Soviet Alliance, 1945–1963, ed. Odd Arne Westad (Washington: Woodrow Wilson Center Press, 1998), 117–40. Orleans estimates that more than 11,000 Chinese received Soviet style training and that 7,324 came back with proper qualifications, while a total of 8,000–10,000 Soviet advisors resided in China from 1950–60 (but only 126 in the years 1950–52). The number is plausible in light of more recent research in Russian archives by Shen Zhihua. According to Shen's findings, the Soviet minister of Foreign Affairs (1949–53) Andrey Vyshinsky (1883–1954) had listed in a secret report (dated April 17, 1952) to the Soviet diplomat-politician and the First Deputy Premier (1942–57) Vyacheslav M. Molotov (1890–1986) similar numbers. See "*Weixinsiji zhi Moluotuofu han: paiqian Sulian zhuanjia wenti* 维辛斯基致莫洛托夫涵：派遣苏联专家问题 (Vyshinsky's letter to Molotov: Issues in dispatching experts from the USSR)," in *Eluosi jiemi dang'an xuanbian—Zhong-Su guanxi* 俄罗斯解密档案选编 中苏关系 (Selected declassified Russian archives: Sino-Soviet relations) (12 vols.), ed. Shen Zhihua 沈志华 (Shanghai: Dongfang chuban zhongxin, 2015), Vol. 4, 212–14. See also Shen's recent study on Soviet experts in China: *Sulian zhuanjia zai Zhongguo* 苏联专家在中国 (Beijing: Xinhua chubanshe, 2009).

10. Kaple, "Soviet Advisors in China in the 1950s," 117–40; Zhou Shangwen 周尚文, Li Peng 李鹏, and Hao Yuqing 郝宇青, *Xin Zhongguo chuqi "liu-Su chao" shilu yu sikao* 新中国初期"留苏潮"实录与思考 (Shanghai: Huadong shifan daxue chubanshe, 2012).

11. On the restructuring of Chinese libraries under the Soviet influence in the 1950s and the ideological consequences thereof, as well as the growth rate of Russian language publication in two major libraries in Beijing, see Priscilla C. Yu, "Leaning to One Side: The Impact of the Cold War on Chinese Library Collections," *Libraries & Culture* 36, no. 1 (2001): 253–66.

12. For a list of these books see Deborah A. Kaple, *Dream of a Red Factory. The Legacy of High Stalinism in China* (Oxford: Oxford University Press, 1994), 14–18.

13. *Lun woguo de kexue gongzuo* 论我国的科学工作 (On our country's scientific work) (Beijing: Renmin chubanshe, 1956), 23. Internal reading materials were texts that were not publicly accessible, but restricted to party cadres and/or members of a distinct institution. They in most cases contained sensitive information, such as reports on other countries or information in fields that were central to national interest.

14. Zhongguo jingji lunwenxuan bianji weiyuanhui 中国经济论文选编辑委员会, *Xiang Sulian zhuanjia xuexi* 向苏联专家学习 (*Learning from the Soviet Experts*) (Beijing: Sanlian shudian, 1953).

15. It is impossible to provide a full list of publications. A good introduction to agricultural mechanization is Alfred L. Chan, *Mao's Crusade: Politics and Policy Implementation in China's Great Leap Forward* (Oxford: Oxford University Press, 2001).

16. Benedict Stavis offers a similar and partly too optimistic picture of China's mechanization, see Benedict Stavis, *The Politics of Agricultural Mechanization in China* (Ithaca: Cornell University Press, 1978).

17. It would be futile to attempt to provide a list of all these manuals and tutorials. The two most prominent and influential publishers during the 1950s and 1960s were

the Agricultural Publishing House (*Nongye chubanshe* 农业出版社, the successor to the publisher *Caizheng jingji* 财政经济 in 1958) and the Science Publishing House (*Kexue chubanshe* 科学出版社, founded in 1954). Alone the list of publications of Agricultural Publishing House includes for the period from 1958 to 1985 more than 4,700 monographs and 27 periodicals. See also Nongye jixiebu bangongting 农业机械部办公厅, *Quanguo nongye jixie shumu 1949–1960* 全国农业机械书目 1949–1960 (Catalogue of Books on Agricultural Machinery in the Whole Country, 1949–1960) (no publisher, 1961).

18. Xu Guohua 许国华, and Liu Jiansheng 刘健生, *Tuolaji he lianhe shougeji* 拖拉机和联合收割机 (Tractors and combine harvesters) (Beijing: Zhonghua quanguo kexue jishu puji xiehui, 1954), 28. The journal *Popular Science* (*Kexue dazhong*) reported in December 1955 that Chinese technicians had succeeded in building a functioning replica of the Stalinez-6, the first Chinese combined harvester. See "Woguo zhizao de diyi tai guwu lianhe shougeji 我国制造的第一台谷物联合收割机 (The first combined grain harvester for grain constructed in our country)," *Kexue dazhong*, no. 12 (1955): 471–72. The Stalinez had originally been built by the Soviet agricultural machinery maker Rostselmasch (Ростсельмаш). Rostselmasch—founded in 1929—started to build combine harvesters in 1931 and named them after Joseph Stalin (1858–1953). In the same year these new harvesters were tested together with machines built by rivaling companies, such as Oliver, Holt, and Caterpillar in the region of Krasnodar. The Soviet model proved to be superior because it was able to harvest also corn, millet, and sunflowers, as Xu and Liu reported in their book.

19. Xu and Liu, *Tuolaji he lianhe shougeji*, 32–33.

20. Mao Zedong, "On the Co-operative Transformation of Agriculture," in *Selected Works of Mao Tse-tung,* Vol. V (Peking: Foreign Languages Press, 1977), 197.

21. Stavis, *The Politics of Agricultural Mechanization in China*, 58–61.

22. Mao Zedong, "Summing-up Speech at Sixth Expanded Plenum of the Seventh Central Committee," in *Miscellany of Mao Tse-Tung Thought (1949–1968)* (Springfield, VA: Joint Publications Research Service, 1974), 16; Mao Zedong, "On the Co-operative Transformation of Agriculture," 197.

23. On the role of Chen and the successes of the Ministry see Chan, *Mao's Crusade*, 109–16. Mao Zedong, "Dangnei tongxin 党内通信 (An Internal Party Letter)," *Jianguo yilai Mao Zedong wengao* 8 (April 29, 1959): 235–40. Mao had called for farm implements research centers already on January 31, 1958, as shows the "Sixty Work Methods (Draft)" (*Gongzuo fangfa liushi tiao* [*caoan*] 工作方法六十条 [草案]), here article 53 reads: "Farm implements research centers should be established in provinces, autonomous regions and municipalities directly under control of the Central Government. They should assume special responsibility in conducting research in improved farm implements and medium and small mechanized farm implements. They should establish close ties with farm implements plans and hand over their research results to the latter for manufacture" (省、自治区、直属市，应当设立农具研究所，专门负责研究各种改良农具和中小型机械农具，同农业制造厂密切联系，研究好了就交付制造), in *Mao Zedong sixiang wansui* 毛泽东思想万岁 (Long live Mao Zedong Thought) (1968), coll. Wang Chaoxing, publ. Gang ersi

Wuhan daxue zongbu 钢二司武汉大学总部; here taken from Stavis, *The Politics of Agricultural Mechanization in China*, 109.

24. "The Conflict Between Mao Tse-tung and Liu Shao-chi over Agricultural Mechanization in Communist China," *Current Scene* 6, no. 17 (1968): 9.

25. Leng Quan 冷泉, "Zenyang genghao de zuzhi guoying nongchang shengchan jingsai? 怎样更好地组织国营农场生产竞赛? (How to better organize production competition in state farms?)," *Zhongguo Nongken*, no. 4 (1953): 13.

26. "Yibu bei diuqi sannianduo de kangbaiyin you dongqi lai le! 一部被丢弃三年多的康拜因又动起来了! (A combine harvester that has been unused for more than three years moves again!)," *Zhongguo Nongken*, no. 9 (1953): 26.

27. Jingji ziliao bianji weiyuanhui 经济资料编辑委员会, *Zenyang tigao tuolaji de gongzuo xiaolü* 怎样提高拖拉机的工作效率 (Beijing: Caizheng jingji chubanshe, 1956).

28. "Tuiguang xinshi chuli nongju 推广新式畜力农具 (Disseminating the latest types of animal-drawn farm tools)," *Renmin Ribao*, January 6, 1955.

29. Nongcun qingnianshe 农村青年社, *Shuanglun shuanghuali jianghua* 双轮双铧犁讲话 (Speech on the two-wheel two-blade plow) (Beijing: Zhongguo qingnian chubanshe, 1956), 6. In addition, the quality of the TWTB plow was low and maintenance service was next to none, as argued by Zhu Xianling 朱显灵 and Hu Huakai 胡化凯, "Shuanglun shuanghuali yu Zhongguo xinshi nongju tuiguang gongzuo 双轮双铧犁与中国新式农具推广工作 (Work to promote the two-wheel two-blade plow and Chinese new farming tools)," *Dangdai Zhongguoshi yanjiu* 16, no. 3 (2009): 56–63.

30. This was the National Programme for Agricultural Production (1956–67), or *Yijiuwuliu nian dao yijiuliuqi nian quanguo nongye fazhan gangyao cao'an* 一九五六年到一九六七年全国农业发展纲要草案.

31. "Yao fandui baoshou zhuyi, ye yao fandui jizao qingxu 要反对保守主义，也要反对急躁情绪 (It is important to oppose conservatism, it is also important to oppose rush emotions)," *Renmin Ribao*, June 20, 1956.

32. Bo Yibo 薄一波, *Ruogan zhongda juece yu shijian de huigu* (*shangjuan*) 若干重大决策与事件的回顾 (上卷) (Reviewing a number of significant strategic decisions and events [vol. 1]) (Beijing: Renmin chubanshe, 1991), 538. Paying an additional 1.4 million work force for constructing 6 million two-wheel and double-blade plows for the necessary mechanization of agriculture as advised in the National Program of Agriculture Development Plan (1956–67) was a huge financial burden. In addition, it was a failure because these plows were too heavy to be pulled by animals and could only be used on large and flat land plots. See Barbara Barnouin and Changgen Yu, *Zhou Enlai: A Political Life* (Hong Kong: Chinese University Press, 2006), 173.

33. On the deep plowing campaign in general see Chan, *Mao's Crusade*.

34. "Tigao shuanglun shuanghuali de xiaoyong 提高双轮双铧犁的效用 (Increasing the effectiveness of the TWTB)," *Renmin Ribao*, April 16, 1958.

35. Nongyebu jixieju 农业部机械局, *Huifu shuanglun shuanghuali de mingyu* 恢复双轮双铧犁的名誉 (Restoring the reputation of the two-wheel, two-blade plow) (Beijing: Nongye chubanshe, 1958).

36. Chan, *Mao's Crusade*, 136.

37. See the report "Zhizhao, tuiguang, shiyong xin nongju—Zhongyang wu danwei zhishi suoshu zuohao nongju gongzuo 制造、推广、使用新农具—中央五单位指示所属作好农具工作 (Manufacturing, disseminating and using new farming tools—five work units of Central Government instruct their affiliated work units to guarantee their work on farming tools)," *Renmin Ribao*, March 20, 1958.

38. "Yao juti lingdao gongzuo gaige—Mao zhuxi yao gedi moqing qingkuang dingchu guihua—Zhonggong Zhongyang shujichu zhaokai dianhua huiyi jinxing tuidong he jiancha 要具体领导工具改革—毛主席要各地摸清情况订出规划—中共中央书记处召开电话会议进行推动和检查 (We have to concretely lead the tool reform—Chairman Mao requests all parts of the country to clarify the situation and draw up plans—The secretariat of the Central Committee of the Communist Party of China convenes a telephone conference on promotion and inspection)," *Renmin Ribao*, August 21, 1958.

39. See "Zao gunzhu zhoucheng yao jianding yikao qunzhong—tiaojian xiangtong, zuofa butong, Xianyang honghong lielie, Xingping lengleng qingqing 造滚珠轴承要坚定依靠群众—条件相同，作法不同，咸阳轰轰烈烈，兴平冷冷清清 (When constructing ball bearing we have to firmly rely on the masses. With conditions being identical constructions methods are not: Xianyang is vigorous, Xingping is desolate)," *Renmin Ribao*, August 19, 1958.

40. See Mao Zedong, "Dangnei tongxin," 244–46.

41. See Jiangsu sheng shuiliting 江苏省水利厅, *Qunzhong chuangzao de watu, hangtu gongju* 群众创造的挖土、夯土工具 (Earth Digging and Ramming Tools Created by the Masses) (Nanjing: Jiangsu renmin chubanshe, 1958); Jiangsu sheng nonglinting nongkenju 江苏省农林厅农垦局 and Jiangsu sheng shougongye guanliju 江苏省手工业管理局, *Nongju gaige congshu: Qunzhong chuangzao de zhonggeng chucao nongju* 农具改革丛书: 群众创造的中耕除草农具 (Book Series on Agricultural Tool Improvement: Plowing and Weeding Tools Created by the Masses) (Nanjing: Jiangsu renmin chubanshe, 1958).

42. The *Serial on Experiences in Revolutionizing Agricultural Technology* (*Dagao nongye jishu geming de jingyan congshu* 大搞农业技术革命的经验丛书) promoted the use of various dissemination techniques, ranging from manuals, practical demonstrations, and exhibitions halls that combined old and new technologies. On the role of exhibitions, see Quanguo nongju zhanlanhui 全国农具展览会, *Nongju tuxuan* 农具图选 (Selected Blueprints of Farming Tools) (vol. 1–20) (Beijing: Nongye chubanshe, 1958); Quanguo nongju zhanlanhui 全国农具展览会, *Quanguo nongju zhanlanhui—tuijian zhanpin—nongtian paiguan jixie* 全国农具展览会—推荐展品—农田排灌机械 (National Exhibition of Agricultural Tools—Recommended Exhibits—Farmland Irrigation Machines) (Beijing: Kexue puji chubanshe, 1958), and especially Shanghai shi nongyeju 上海市农业局, *Shanghai shi nongye zhanlanhui* 上海市农业展览会 (Agricultural Exhibition in Shanghai) (Shanghai, 1957).

43. This was made clear in a number of publications such as Wang Jiankun 王建坤 and Zhang Qinghua 张清华, *Gailiang nongju de qiaomen* 改良农具的窍门 (*The Know-how of Improving Farming Tools*) (Beijing: Zhongguo qingnian chubanshe, 1958); Jiangxi sheng nongyeting 江西省农业厅, *Gailiang nongju de zhizao yu*

shiyong 改良农具的制造与使用 (*Construction and Use of Improved Farming Tools*) (Nanchang: Jiangxi renmin chubanshe, 1955); *Gailiang nongju jieshao* 改良农具介绍 (*Introducing Improvements of Farming Tools*) (no publisher, 1956); Zhonggong Shanghai shiwei nongcun gongzuo weiyuanhui nongju gaige bangongshi 中共上海市委农村工作委员会农具改革办公室, *Shanghai shi xinshi nongju xuanji* 上海市新式农具选辑 (Selection of New Agricultural Tools in Shanghai) (Shanghai: Kexue jishu chubanshe, 1960).

44. Mao Zedong, "Dangnei tongxin," 244–46.

45. Anna Louise Strong, *The Rise of the People's Communes in China* (New York: Marzani & Munsell, 1960), 41–42.

46. For this increase see also the assessement in "Dagao gongju gaige 大搞工具改革 (Enthusiastically setting up tool improvement)," *Renmin Ribao*, August 21, 1958.

47. *Yong tu banfa shixian ban jixiehua* 用土办法实现半机械化 (Using local methods to achieve half-mechanization) (Nanjing: Jiangsu renmin chubanshe, 1958), 7–9.

48. Quanguo nongju zhanlanhui, *Nongju tuxuan*.

49. Quanguo nongju zhanlanhui, *Quanguo nongju zhanlanhui—tuijian zhanpin—nongtian paiguan jixie*, no. 146. In the same year, an article appeared in the journal *Zhongguo Nongken* that praised this wheelbarrow and introduced it as a precious contribution of Henan's peasants to the modernization of Chinese agricultural production. See Ni Xiulian 倪修莲, "Henan nongmin chuangzao de jizhong tufang yunshu gongju 河南农民创造的几种土方运输工具 (Henan's peasants create several kinds of earth transportation tools)," *Zhongguo Nongken*, no. 5 (1958): 23–25.

50. Quanguo nongju zhanlanhui, *Quanguo nongju zhanlanhui—tuijian zhanpin—nongtian paiguan jixie*, no. 146.

In addition, the chart shows also how many wood boards, nails, and iron wire are necessary for construction, and how many labor hours are needed. This follows a simple logic that is present in any economic system, namely cost calculation. The same table and a similar description of this wheelbarrow can also be found in Wang Jiankun and Zhang Qinghua, *Gailiang nongju de qiaomen*.

51. Thus, also the insistence of the peasants to have a son as offspring. In the countryside, this is rather an economic question, and less one of culture, because a daughter could hardly contribute to physical labor in the same amount as a healthy son. Children and young people thus participated actively in labor, similar to the situation in early modern Europe. See Philippe Ariès, *Centuries of Childhood: A Social History of Family Life* (New York: Alfred A. Knopf, 1962).

52. "Funü yuejinche 妇女跃进车 (Women's Leap Forward Handcart)," in Shaanxi sheng nongye gongju gaige zhanlanguan 陕西省农业工具改革展览馆, *Xianjin gongju tupu* 先进工具图谱 (Atlas of Advanced Tools) (Xi'an: Shaanxi renmin chubanshe, 1960), 170. A water pump driven by seesaw technique (*Erlong xizhu shuiche* 二龙戏珠水车) that can be operated by women and children is also introduced in this book, see p. 53.

53. *A Handcart with Steel Balls* (*Gangzhu feiche* 钢珠飞车), directed by Zeng Weizhi 曾未之 (1959).

54. Lin Shan 林山, "Jixu guzu ganjin, shixian jinnian nongye kexue gengda de yuejin! 继续鼓足干劲，实现今年农业科学更大的跃进! (Continue to do one's best to realize this year's even bigger leap forward in agricultural science)," *Guangxi nongye kexue* 12 (1959): 1–4, 29.

55. Harding, Sandra, *Sciences from Below: Feminisms, Postcolonialities, and Modernities* (Durham: Duke University Press, 2008).

56. Fa-ti Fan, "Science, State, and Citizens: Notes from Another Shore," *Osiris* 27, no. 1, Clio Meets Science: The Challenges of History (2012): 242.

56. Schmalzer, *Red Revolution, Green Revolution*, 107.

Chapter Four

Producing Knowledge on the Shopfloor

Technological Innovation in Socialist Industrialization

In the Mao era, the production of steel and iron had been viewed as the leading indicator of China's level of modernization and the basis of realizing its agricultural mechanization. Therefore, the metallurgy industry had been subject to the state's direct interventions. Since the 1950s it had experienced and hence testified to the most violent vacillation of state policies concerning knowledge production. In this field, the notion of "science" was perceived primarily in terms of "technology," which measures the validity of knowledge by means of its application on the shopfloor. The constant search for new ways of (the Party) organizing such knowledge production in the Mao era attempted a synthesis of expert knowledge and knowledge created by the industrial worker. Embedded in the negotiations between the two were the political agenda of achieving national independence and self-reliance through industrial modernization and the social ideal of leveling classes by cultivating a new working class.[1]

As discussed in the preceding chapters, the early PRC enthusiastically pursued the socialist modernity exemplified by the self-portrait of the Soviet Union. This chapter traces China's adoption of and then movement away from the Soviet model in search of its own, pragmatic path to industrialization. It did so by constantly reorganizing knowledge production on the shopfloor to deal with resource constraints and to forge the consciousness and identity of the industrial worker as an educated, progressive social class. The first section analyzes China's adoption of industrial production and management of High Stalinism in the early 1950s, which highlights social agendas and the Party's role in knowledge and economic productions.[2] The model worker Wang Chonglun 王崇伦 (1927–2002) from the heavy industry sector, in particular, embodies the ideal of the emancipated worker developing into the creative knowledge producer on the shopfloor under the leadership of the Party. His technological innovations were disseminated so as to routinize his individual "miraculous" achievements into collective productivity. In the fanatic pursuit of accelerated modernization during the Great Leap Forward years, the second section shows that non-contingent experiment supported by numbers was deployed as an empirical scientific approach in public communication, which justified the replacement or making of the steel with indigenous products such as nodular cast iron, so-called "native iron" (*tutie* 土铁, or indigenously made pig iron), and "white steel" (ceramic products used in place of steel under certain circumstances). The last section conducts a case study of the Anshan Iron and Steel Works (hereafter its abbreviation the AnSteel), which had been the flagship of China's socialist industrialization. Delineating its changing role in defining approaches to and authority of knowledge in the field of metallurgy, this case study demonstrates the interactions between the AnSteel and the state's policymaking, which vacillated violently in the leadership's struggles to deal with polito-socio-economic issues in the turbulent years of the 1950s to the 1970s.

SOCIALIZING KNOWLEDGE AND INDUSTRIAL PRODUCTION

The new China hoped to initiate a bottom-up surge of scientific knowledge production based on the first-hand experience of the masses. In industry, the factory shopfloor functioned not only as the location of economic production but also as the venue where knowledge was created and disseminated. The early PRC adopted the Soviet model of High Stalinism in order to industrialize the country and create a Chinese urban proletariat. Deborah Kaple shows that the transplantation of the Soviet model into the PRC between 1949 and 1953 held a particular interest in labor force development and management.[3] Among many things adopted and adapted from the Soviet model in the early 1950s, four features—which themselves were from different periods of Soviet economic production—should merit our attention, because they are particularly important for our examination of the dynamics among the Party, the management, the worker on the ground, and economic productivity: 1. the installation of the Party organization at all levels of industrial enterprises; 2. the debate over the managing principle of one-man management, or *edinonachalie* (единоначалие *yizhang zhi* 一长制); 3. various forms of production competition (*shengchan jingsai* 生产竞赛) among workers; and last but not least, 4. the socialist distribution of remuneration.

Two decades after its foundation, the CCP reestablished its identity as a proletarian party by following the Soviet model in industrial organization.

The *Book Series of Studying Soviet Experience of (Economic) Construction* (*Sulian jianshe jingyan yanjiu congshu* 苏联建设经验研究丛书) discussed the role of the Party in industrial production as the organizer, the agitator, and the supervisor. The Party whose branches should be built up all the way to the basic level of working groups, as the Party secretary of Krasnyi Proletarii (Red Proletariat) Moscow Machine-Tool Plant G. Tsarev elaborated in his book, had the responsibilities of inspiring the masses for production, helping the managers to organize human and material resources, and overseeing the implementation of the production plan.[4]

The managerial principle of one-man management promoted during Stalin's First Five-Year Plan (1928–32) "sought to establish strong and efficient management" in the industry sector and, by enhancing both managerial authority and accountability of the Communist (red) managers, expected them to combine the functions of the Party and professional work in the factories.[5] Kuromiya's research points out, however, that the majority of red managers in the early 1930s could not live up to "their powers and responsibilities."[6] In 1935, *edinonachalie* was replaced by the Stakhanovite movement, a mass campaign form of socialist competition named after the miner Alekseĭ G. Stakhanov (1906–77) for his extraordinary efficiency in production.[7]

Edinonachalie was introduced into the PRC in the early 1950s, but it was not until 1953—the beginning of the First Five-Year Plan—that a campaign was launched to implement it in the whole country. In comparison with the political concern of "bourgeois specialists" in the Soviet Union, the Chinese interpretation prioritized the manager's professional knowledge of economic production as the prerequisite. In a thin 1954 book on how to implement the one-man management in industrial and mining enterprises, the author justified this centralized leadership and its authority by stressing the manager's professional knowledge about the whole production process and technical issues, which could guarantee the proper organization and smooth execution of economic production.[8] William Brugger believes that one-man management was only implemented partially in China and over a short period of time between 1953 and 1956.[9] Nevertheless the explicit endorsement and empowerment of the expert and his professional knowledge would turn this managerial principle into the target attacked by Mao Zedong in 1960, when he championed the so-called AnSteel Charter (*Angang xianfa* 鞍钢宪法) to promote a presumably more democratic and efficient—mass-line, non-professional, bottom-up—approach to knowledge production in industry.

The early 1950s also saw various forms of mass production competition in China as socialist emulation (*shehuizhuyi jingsai* 社会主义竞赛), following the Soviet examples of shock workers and Stakhanovites who "fulfilled obligations over and above their planned quotas."[10] These production competitions assumed different forms and names, addressing various political and economic concerns at the time. The Patriotic Production Competition (*Aiguozhuyi shengchan jingsai* 爱国主义生产竞赛) in 1950 and 1951 took place against the political background of the Korean War. Its initiator Zhao Guoyou 赵国有 (born 1924), a model worker of the Third Machinery Factory in the Northeast, requested in October 1950 that workers "turn the factory into the battlefield and the machine into the weapon."[11] The Competition of Increasing Production with Less Cost (*Zengchan jieyue jingsai* 增产节约竞赛) in 1952 was carried out to improve both the productivity and political consciousness of workers. The nation-wide Campaign of Technological Innovations (*Jishu gexin yundong* 技术革新运动) in 1954 encouraged tool improvement as well as new techniques and technologies to accelerate industrialization with no extra or even less cost.

Stalin differentiated the competition in capitalism and emulation in socialism in his review of E. Mikulina's pamphlet *Emulation of the Masses* (1929), arguing that the latter is to "catch up with the best and secure the advance of all."[12] The idealized picture of socialist emulation therefore portrayed the best and the most productive model workers emerging as emancipated laborer with tremendous initiative and creativity, who led his colleagues to achieve what he had done. Obsessed with numbers quantifying their productivity, these competitions linked numerically measured productivity with the social process of building working-class identity and political commitment. Similarly, socialist emulation in the Chinese context was seen as a means of developing new work ethics. Embodied in the model worker, the new work ethics should help others to grasp their new identity as the owner of the factory and as the master of the new country. The problems and conflicts of production competition in the workplace, especially the imposed extra workload and social pressure of isolating "unproductive" workers, were depicted in a 1949 four-act play *Red Flag Song* (*Hongqi ge* 红旗歌).[13]

The early PRC also imported from the Soviet Union the socialist distribution of remuneration, that is, the worker "must be remunerated for his special skills, his efficiency on the job, and according to the quality and quantity of his output."[14] Such a piece-rate payment system means that model workers of the early PRC, similar to shock workers and Stakhanovites in the Soviet Union, were rewarded both materially and politically. This would be canceled during the Great Leap Forward years and denounced in the Cultural Revolution, when the worker was expected to derive their enthusiasm and initiative solely from their "communist consciousness." Yet monetary incentive was an unspoken part of the motivations during the Mao era.

One Chinese version of the Stakhanovite was the worker-innovator Wang Chonglun of the General Machinery Works of the AnSteel in Liaoning Province. Supported by the Party, he was able to apply new techniques and technological innovations to increase labor productivity in his workplace. Therefore, he embodied the new knowledge producer envisioned by the Party, who contributed to the idea of "science and technology" with their practical experience on the shopfloor while blurring the boundaries between manual and mental work.[15] In a word, Wang stood for the creativity, enthusiasm, and high productivity of the emancipated worker in a new socialist country, and even more important, his experience was learnable and hence could be disseminated and routinized.

Among many of Wang's inventions and innovations, the most well-known was the "universal fixture" (*wanneng gongju tai* 万能工具胎), an auxiliary tool that could be attached to a shaper to make it function like a slotting machine. This invention had reportedly streamlined the producing process in his work unit tremendously: it not only solved the problem of slotting machine shortage in Wang's workshop, but could also be used to produce fifteen different parts for rock drills. With his improved and invented tools, Wang was able to finish his production quota four years before the plan in 1953.

The story of Wang Chonglun and his innovations is a narrative of the knowledge producer on the shopfloor whose staggering productivity was made possible by the Party and his own growing class consciousness. The poet Wei Wei 魏巍 (1920–2008) claimed that Wang was a man "walking ahead of time," representing the zeitgeist that "ordinary people" were now able to produce "miracles."[16] Wang's own essay in the professional management journal *Newsletter of Heavy Industry* (*Zhonggongye tongxun* 重工业通讯) claimed that his initiative and enthusiasm for innovation and invention were inspired by the elevated social status of the manual worker.[17] In his autobiographical narrative, he enumerated all sorts of support from the Party secretary Bai Mingxin 白明新, including persuading other workers to help Wang in production conferences, organizing necessary parts for his technological innovations, and allowing him to do experiments during working time.[18] The story of Wang Chonglun is therefore one about the integration of the Party in the economic process, which enabled the transformation of the ordinary young worker into an avid innovator proud of his new working-class identity and eager to realize the country's accelerated modernization.

Machine Workers (*Jixie gongren* 机械工人), a popular science magazine targeting factory workers, allotted three pages to the blueprint, operating instructions, and technological advantages of the universal fixture.[19] It further rendered Wang's amazing efficiency learnable by attributing it to his good working habits and modesty, such as his meticulous maintenance of tools and machines, adherence to proper operating procedures, his exact use of time as well as his humbleness to seek support from the Party, technicians, and skilled workers.[20]

His innovations were disseminated by means of demonstration on various national exhibitions and in workshops. Following the example of Stakhanovite training schools in the Soviet Union, a short-term school was set up by Wang Chonglun's work unit in his name in early 1953. Fifty-one machine workers selected from various units came to watch his demonstration. They would then disseminate what they had learned to their own work units.[21] In this school not only Wang himself lectured, but other innovators also had the opportunity to present their achievements. Staged on the shopfloor, these demonstrations as a form of knowledge transmission were boosted by technical diagrams, political slogans, and blackboard propaganda. A trained guide should explain the innovation or invention before the demonstration, clarifying the inventor, the time of invention, and comparing its productivity before and after.[22] The promotion of the advanced experience was not so smooth in practice, though. Workers were suspicious of the miraculous productivity or found Wang's working conditions too good to be applicable to their own factories.[23] Nevertheless his achievements reportedly fanned out all over the country: his own workshop had finished the planned production quota one year ahead; the workers of the General Machinery Works and other industrial enterprises were actively improving their tools and facilities; workers from other economic sectors swore to learn from Wang to complete or over-fulfill their production plan, the *People's Daily* reported enthusiastically.[24]

In 1956, the sustainability of experience-based, empirical knowledge production to realize further technological progress was put into question. As mentioned in the first chapter, Zhou Enlai's 1956 report "On the Problem of Intellectuals" shifted the emphasis onto theoretical science and professional knowledge. What had remained consistent was the goal of accelerating national industrialization and the mass-line policy of knowledge production.

PRESENTING TECHNOLOGICAL INNOVATIONS

Zhou Enlai's report restored precariously the authority of natural scientists in their trained areas and encouraged them to work together with peasants and workers. In June 1958, the two editorials calling respectively for the "revolution in technology" and the "revolution in culture" strived to realize a "cultural turnover" (*wenhua fanshen yundong* 文化翻身运动) of all manual laborers—from their ignorance and awe of professional knowledge—and to train "a ten million strong army of working-class intellectuals (*gongren jieji zhishifenzi* 工人阶级知识分子)" in ten to fifteen years. The editorials stated that science and technology and human civilization would accelerate drastically if laborers "master knowledge and are able to combine closely theory and practice of production (*shengchan shijian*)."²⁵ These editorials therefore not only reiterated Mao Zedong's epistemology of practice as a valid approach to knowledge but also built it into the class-leveling agenda.²⁶ In this section, we study three cases of technological innovation that followed the principles of "more, faster, better, and more economical" during the Great Leap Forward years. We select these cases from major official media and professional journals—for example, the *Journal of Metallurgy* (*Yejinbao* 冶金报), *Newsletter of Heavy Industry*, and *Steel and Iron* (*Gangtie* 钢铁)—as well as in popular science magazines, booklets, and pamphlets. It is notable, first, that they foreground the important role of both professional technocrats and researchers and the masses in knowledge production; and second, that experiments and numbers involved in these cases were presented mainly to persuade non-experts, which downplayed and even concealed the contingency and controversy in the process of knowledge production.

In the pamphlet *The Foundation of Industry—the Iron and Steel Industry* (*Gongye de jichu—gangtie gongye* 工业的基础—钢铁工业), Mao Zedong and Stalin were quoted on the first pages to illustrate the significance of industry to nation-building: "Steel and iron are the foundation of industry" and "without industry, there will not be solid national defense, the well-being of the people, and a wealthy and strong nation."²⁷ The new state was keenly aware of the central role of steel for creating a new modern China and it was impatient with the cost and time to develop the iron and steel industry.

There is no more revealing example than that of nodular cast iron to demonstrate the country's urgent desire to produce iron and steel for a fast industrialization. Also known as ductile iron, nodular cast iron was first produced in laboratories in Germany and Britain in 1939. In the 1940s the American metallurgical engineer Keith Millis (1915–92) made it possible for industrial production and had it patented in 1949.²⁸ The Research Department of Chinese Academy of Science introduced nodular cast iron in the *People's Pictorial* (*Renmin Huabao* 人民画报) in 1952, declaring that they had successfully produced it in October 1950. The particularity of this product lies in its much higher tensile strength than that of normal cast iron, which was illustrated in the microstructure photos comparing its nodular-shaped graphite with the flake-formed graphite of normal cast iron. According to this report, nodular cast iron could be produced in ordinary iron foundry works, cost less yet possessed similar tensile strength to that of cast steel, thus it was welcomed as a solution to the serious shortage of high-quality steel in the fast development of China's industry. Since 1951, the article stated, nodular cast iron had been used in many areas, including textile machine spare parts, cylinder casing of automobiles as well as agricultural tools such as the plow.²⁹ In a word, although nodular cast iron was not the same as cast steel, it required simpler cast facilities, less investment, easier ingredient control, and therefore was a good ersatz to overcome steel shortages.

The *People's Daily* reported in September 1958 on a successful experiment of manufacturing railway tracks using nodular cast iron. Carried out by Tsinghua University students and an "old master worker" (*lao shifu* 老师傅) with assistance from Shijingshan Iron and Steel Works in Beijing, this experiment reportedly demonstrated the possibility of using nodular cast iron to make railway tracks for medium- and small-sized factories and counties.³⁰ This reported case anticipated some major arguments promoting nodular cast iron in the Great Leap Forward years which, on the one hand, represented the ideal of combining expert/theoretical knowledge with the knowledge gained on the shopfloor, and on the other hand, negotiated between conventional technical standards and pragmatic purposes.

Toward the end of the year 1958, a series of articles were published in the popular science magazine *Knowledge Is Power*, calling for disseminating the manufacturing methods and applications of nodular cast iron in industry and infrastructure construction. Two articles by technocrats—the vice minister of the Ministry of First Machine Building Industry Wang Daohan 汪道涵 (1915–2005) and the vice minister of the Ministry of Railways Yu Guangsheng 余光生 (1907–78)—set the keynote that they were both technological exploration and "a political task."³¹ Five popular science writings by the Teaching and Research Group of Casting Technology at Tsinghua University (*Qinghua daxue zhugong jiaoyanzu* 清华大学铸工教研组) introduced the new material nodular cast iron, its nodulizing process, a new nodulizing element, an "indigenous method" (*tu banfa* 土办法) of saving magnesium—the expensive nodulizing element—and how to manufacture railway tracks from nodular cast iron.³² The last two articles—one by the railway track research group affiliated to the Ministry of First Machine Building Industry and the other by a researcher from the Research Institute of Machine Building and Techniques—proposed replacing steel with gray iron to construct a so-called "indigenously-made railway" (*tu tielu* 土铁路) for less-traveled routes and even nodulizing "native iron" into "native" nodular cast iron (*"tu" qiumo zhutie* "土"球墨铸铁).³³

Wang Daohan promoted nodular cast iron as a solution for the machine building industry and argued for its use in place of cast steel and forged steel for manufacturing *some* (*yibufen* 一部分) machine components. Attributing the "miracles" of nodular cast iron to the wisdom of the masses and the glorious leadership of the Party, he called for accepting nodular cast iron as a proper and non-makeshift material and urged for further experiments to search for solutions to existing problems, including finding ways to produce nodular cast iron from "native iron." He nevertheless pointed out that the use of nodular cast iron in *each* (*mei yizhong* 每一种) product should be evaluated precisely in their concrete economic and technological contexts. Following the guidelines in Wang's article, Yu Guangsheng advocated for "walking on two legs" as a quick solution to the shortage of steel tracks in developing China's railway transportation system. In other words, steel tracks produced by large modern steelmaking works should be supplemented by those developed indigenously and produced from other materials, which Yu claimed would function well for less busy branch lines and local lines. Thus, both Wang and Yu perceived the use of nodular cast iron as a way of adapting technology to local needs and, by emphasizing the need of evaluating each product in context, implicitly conceded its nature of experiment.

The experiments and inventions of the Teaching and Research Group of Casting Technology at Tsinghua University, which were introduced in the following pages in the same issue of the magazine, valorized Wang Daohan's statement that "science and technology can develop fast only when they serve (economic) production."[34] They claimed that their "indigenous method" not only helped increase the safety and precision of adding the nodulizing element magnesium but also reduced the required amount from the normal percentage of 0.4 percent to a mere 0.07 percent. Nodular cast iron railway tracks, on the other hand, were a result of negotiations between technical criteria and various priorities of economic production, for example, the research group processed nodular cast iron with heat treatment in order to heighten its elongation—at the cost of its tensile strength—to satisfy the needs of the railway track. They also mentioned the difficulty of quality control and the efficiency problem in producing these tracks.

The problems in making and adopting nodular cast iron for local needs of economic production were mentioned but brushed away in the last two articles, which pushed even more radical ways of building railway tracks. One of them suggested using gray iron to build railway tracks for local lines and slow trains. Hardly supported by any technical data, this article built its argument on the low cost of gray iron and a comparison of it with even more brittle white iron. It conceded that gray iron has notoriously weak impact resistance, but strived to argue that the improvement of casting technology (in the future?) and the reduced speed of trains would make the proposal feasible.[35] The other one argued for the feasibility of using "native iron," which often contains a high percentage of sulfur and other impurities, to make "native" nodular cast iron. The content of sulfur, so went the smooth theoretical reasoning of the author, may contribute to the nodulizing process and save on magnesium. This article drastically downplayed a series of technical and logistical difficulties, for example, the technical control of using sulfur in the intended way; quality control of native iron, whose metallic composition depends heavily on the location and technology of its production, etc.

The authorship of these articles, ranging from high-ranking technocrats and research institutes to teachers and students of the elite Tsinghua University, indicates that manufacturing nodular cast iron belonged to the area of the expert. Yet the Tsinghua group image visualized on the cover of *Knowledge Is Power*—a young man with the school pin of Tsinghua University on his chest—emphasized the image of a student-worker on the shopfloor, that is, a working-class intellectual, who presented a piece of track made from nodular cast iron as the result of knowledge from—and for—practices of economic production.[36]

These articles, with their arguments on using nodular cast iron to replace steel under specific circumstances, crystalize the negotiations among technical criteria, material shortage, priority of economic development as well as the social agenda of eliminating the differences between the mental and manual labor. One can easily detect the increasing radicalization of arguments when they moved from the central government level to the local. In

Figure 4.1. Cover page, *Knowledge Is Power* (*Zhishi jiushi liliang* 知识就是力量), no. 12 (1958).

this process, the technological complexity and contingency involved in the reported scientific experiments became increasingly invisible. As the result, the boundaries between the prospective technological innovation and the adopted technology, the feasibility in theory and that in practice were all blurred. Although published in one issue, these articles in a way show how the discrepancy between a national policy and its local (radical) implementations emerged in those years.

Nowadays the production of "native iron" and its use for making steel have already become a notorious symbol of the Great Leap Forward mania of ignorance and voluntarism.[37] Yet the case of "native iron" shows that the justification of its use in metallurgy followed certain "scientific" procedures, including the endorsement of professional journals, experiment as well as statistical numbers as "a tool of persuasion and a basis for rational, methodical, calibrated, and repeatable actions."[38] Although these scientific methods skewed in practice, they nevertheless confirmed the authority of "science," as shown in the *Journal of Metallurgy*.

This official journal of the Ministry of Metallurgy demanded in October 1958 that "native iron can make steel and has to be able to make steel," which was "a political task as well as a technological task." It recommended measures for producing steel from native iron, though conceding that the normal standard for steelmaking was hard to achieve. Native iron often contains high percentage of impurities—sulfur, phosphorus, silica, etc.—that contaminate the high-quality iron and steel in the smelting process.[39] In addition, quality control of native iron was extremely difficult. The Chinese standard of maximum permissible sulfur content in all steel at the time was 0.055–0.06 percent, which was already laxer than the international standard of 0.02–0.04 percent. This standard was further compromised to 0.08–0.12 percent for steel used for manufacturing daily utensils, agricultural tools, and metal hardware for house-building, etc.[40]

A successful story of making steel from native iron in a 1.5-ton basic steel converter came from the experiments conducted by a research group affiliated with the Ministry of Metallurgy and the workers in Xinxing Steelworks in Tianjin.[41] Their experiments were repeated for demonstration and promoted in an onsite conference in the steelworks. *Science Pictorial* published the key points of the conference, which included skills and tips of categorizing native iron of different quality and detailed operation instructions.[42] *The Journal of Metallurgy* claimed that 95 percent of the steel made from native iron in Xinxing Steelworks reached the national standard.[43] An article published in *Machine Worker,* which provided more technical data, reveals that the sulfur content in the steel made from native iron by Xinxing Steelworks could only pass the new standard of 0.08–0.12 percent.[44] In other words, the steel would have been defective had it been measured by the earlier national standard. In 1959, an exhibition of tools made from native iron and native steel was held by the Ministry of Metallurgy and Industry showcasing their usability for making agricultural tools and therefore contributing to "the industrialization of the commune and agricultural mechanization."[45] Here the compromised standard was completely eclipsed by the fanfare of the empirical approach of experiment and onsite conference that disseminated advanced experiences.

If native iron is indigenous because it used local materials and self-made facilities, then the indigenousness of "white steel" was rooted in the Chinese traditional craft of ceramics. The idea was to use ceramic, which cost less to produce, to replace steel to make machines and mechanical components that shared similar qualities, such as hardness, heat resistance, corrosion resistance, low thermal expansion, insulation, and anti-oxidation. Backed up by pictures of ceramic ball-bearings, pumps, and blast blowers for a self-made blast furnace, a 1958 article in *Science Pictorial* claimed that the science of ceramics had developed to such a level that they could be used in place of steel in some areas of machine-building. While it enthusiastically celebrated the advent of the era of the cermet,[46] the article actually focused on ceramic and had to concede that ceramic was brittle and its compressive strength and thermal conductivity could not be compared to steel.[47] According to a researcher of the Institute of Ceramics, it was not yet foreseeable that its features could be changed so much that it could become similar to those of steel and iron. "To rejuvenate ceramics," as the researcher put it, was to improve the qualities of the material and optimize its structure so that it could replace steel to make machines and facilities *under certain circumstances* (*zai yiding de tiaojian xia* 在一定的条件下).[48] In the *People's Daily*, the cautious tone of researchers was replaced with the enthusiastic celebration of a significant technological innovation, which showed that "China's ceramics industry with a thousand-year glorious history has now entered the new era of mechanical ceramics."[49] In 1962 such "white steel" remained a fantasy in a science fiction story.[50] It was not until 1992 that the invention of a ceramic roller was reported in *People's Daily*, inconspicuously in two sentences.[51]

It should be noted that many machines and mechanical components made from "white steel," like those made from native steel produced in Tianjin Xinxing Steelworks, were designed primarily for rural use: ceramic pumps for irrigation, ceramic ball-bearings to be used in transportation vehicles, and even ceramic transportation railway tracks for rural use only.[52] All these repeatedly testified to the state's limited investment in the countryside, and consequently, its expectation of indigenous methods for local tool improvement—as we have argued in the last chapter.

DEFINING AUTHORITY OF KNOWLEDGE PRODUCTION

Reputed as the "eldest son of the (People's) Republic," the AnSteel had completed three major infrastructure projects—the No. 7 blast furnace, the seamless tubing mill, and the steel rolling mill—in 1953 and thereby become *the* flagship of socialist industrialization throughout the Mao era. What has often been left out of its story is the fact that it was originally built in 1918 as a subsidiary of Japan's South Manchurian Railway Company. In 1933, it became a separate Japanese colonial company and from 1936 on had served the needs of the war program against the United States and United Kingdom, notorious for its inhumane working conditions. It resumed production after the CCP's takeover in November 1948. Between 1950 and 1954, tremendous efforts were made to bring the productivity of its main sectors back to their pre-war peak time level.[53] The leading status of the AnSteel in the making of technological knowledge and economic production in the Mao era have been illustrated in a plethora of propaganda sources, including news reports and photos, popular science pamphlets and magazines, exhibitions, documentary films, etc. It had been profoundly involved in defining the authority of knowledge production then.

Matsumoto attributes the AnSteel's "miraculous" restoration of production in the early PRC years to its ample human resources, captured technical documents, and use of the remaining equipment. Young engineers and workers were summoned from all over China; engineers of Japanese origin and the Nationalist past as well as remaining Chinese skilled workers were employed to help to nurture and convey the know-how and skills. This relatively flexible policy of human resources, Matsumoto posits, worked hand in hand with an established training system and the exceptional enthusiasm of its employees to modernize China. Workers and engineers without the Nationalist background were offered opportunities of training and promotion during the period of restoration and rebuilding, a fact confirmed by other sources, which show that activists and model workers were often promoted to management positions in heavy industry working units.[54] Another key factor contributing to the AnSteel's fast recovery was the assistance of Soviet experts, who introduced systematically industrial management patterns, including the abovementioned one-man management and the centralization of production processes such as an internal accounting system, procurement of supplies and selling of outputs, etc.[55] By 1954 the AnSteel had become a symbol of socialist industrialization associated with the most advanced modern technology in metallurgy, industrial management, socialist friendship with the Soviet Union, and a progressive working class with the skills and fervor to achieve China's economic self-reliance.

In April 1954, the Exhibition of the AnSteel Technological Innovations was held in Beijing, displaying its achievements in infrastructure construction and technological innovation. The more important message of this exhibition, as the editorial of the *People's Daily* pointed out, was to gear the emphasis of production competition from shock work to new technologies and techniques. The AnSteel's technological innovations should attest to the great potential of mobilized workers, with its emphasis on improving the efficiency of old and present facilities and optimizing productive procedures.[56] These goals imply the constrained financial resources at the disposal of China's modernization project, which explains why pragmatism had been a central theme during the Mao era. In support of the *People's Daily* editorial and the mass campaign, a documentary titled *The AnSteel under Construction* (*Angang zai jianshezhong* 鞍钢在建设中) was released in 1954 by the Central News Documentary Studio to showcase the successful infrastructure construction of the AnSteel. Various publications appeared discussing how to adopt its experience of technological innovations.[57]

These published sources acknowledged the authority of professional knowledge of Soviet experts, Chinese technicians, and skilled workers on the ground. The *Newsletter of Heavy Industry* devoted much space of its thirteenth issue of 1954 to a series of articles expounding the AnSteel experiences. The chief engineer of the steel mill Ma Chengde 马成德 (1919–2003) stressed the importance of learning from Soviet experts.[58] The Ansteel identified the following five sources essential to innovation: 1. advice from Soviet experts and foreign new technology; 2. rationalization proposals made by workers, technicians, and managerial members; 3. advanced working techniques developed by workers and technicians; 4. advanced experience from other Chinese metallurgical enterprises; and 5. research results within the AnSteel and of scientific research institutes.[59] These articles endorsed the authority of the expert and professional knowledge, which were represented by Soviet experts and research institutes in the AnSteel, while at the same time recognized various knowledge producers such as experts and researchers, technicians, and workers on the shopfloor as well as managerial members.

The 1956 mass campaign of "marching toward science" again confirmed and highlighted the authority of professional knowledge and its effective application. Four short reportage narratives in the book titled *On Their Way Marching towards Science* (*Zai kexue jinjun de daolu shang* 在科学进军的道路上) set the AnSteel as the background to portray a picture of young technicians and workers balancing their theoretical training and shop floor work experience. There young people—students, technicians, and workers—received professional trainings at various levels to become red specialists.

In these narratives the source and authority of knowledge were multiple: old scientists, Soviet experts, and skilled workers; while valid knowledge, in addition to first-hand experience on the shopfloor, also included a whole set of research processes, ranging from collecting reference materials, designing and discussing field research results and experiments with colleagues to writing up lucid scientific papers and presenting their findings to the expert community.[60] Here the steel mill was depicted as an idealized environment of learning, where an open approach to knowledge production was valorized and the aspirations of the workers to become experts were endorsed. And the Spare-time Industrial University of the AnSteel (*Angang yeyu gongye daxue* 鞍钢业余工业大学), founded in 1953, seemed to have offered perfect opportunities for young worker to learn. Yet inside information submitted to high-ranking leaders shows that the school suffered the lack of teaching staff and many registered worker-students quit due to the physical stress of having to study after work and the clumsy administration that failed to facilitate their spare-time learning.[61]

At the beginning of the Great Leap Forward campaign, the AnSteel, whose reputation was mainly founded on its modern industrial technology, could hardly produce sensational news among the mushrooming reports on "indigenous" methods of producing iron and steel. Instead it suffered food shortages already in 1957, which would lead to the deterioration of the workers' physical power and cause more accidents, and then the shortage of raw materials and power failure resulted in the reduced production and shutdown of several rolling mills. By the end of 1958, most units of the AnSteel canceled the piece-rate salary as an indication that the workers had developed "communist consciousness." This change led to the decline of wages of workers, in particular that of skilled workers.[62]

Despite all these problems, the Party committee of the municipal government of Anshan still managed to submit a report to Mao Zedong in March 1960, claiming the success of their technological revolution as part of the Great Leap Forward campaign.[63] Mao's commentary on this report enthusiastically endorsed these principles and celebrated them as the AnSteel Charter. The neologism was coined to contrast the Soviet way, what he named the MaSteel Charter (*Magang xianfa* 马钢宪法), referring to Stalin's showcase enterprise of Soviet industry, Magnitogorsk Iron and Steel Works (MMK).[64] Plenty of studies have shown that the patterns of management and production in the MMK were complex and far from consistent,[65] therefore the so-called MaSteel Charter in Mao's comments should be understood as his strategy of oversimplifying the attack target. In his comments Mao encouraged further technological revolution, mass campaign, politics in command, director-responsibility under the leadership of the Party, and most famously, the so-called "Two Participations, One Reform, and Three-in-one" (*liangcan yigai sanjiehe* 两参一改三结合) policy, namely, the participation in production by managerial members and the participation in management by ordinary workers, the reform of unreasonable rules and regulations, and the collaborations of the worker, the technician, and the managerial member. In other words, the naming of the AnSteel Charter conveyed Mao's explicit message to give up the Soviet model of modernization for an indigenous—or self-reliant—way of modernization. By emphasizing the "three-in-one" dynamics in producing technological knowledge, the AnSteel Charter reaffirmed the mass-line approach to knowledge on the shopfloor, reinstated the Maoist epistemological notion of "practice," and accentuated the political potential of flattening hierarchy within the production process. With the AnSteel Charter, the showcase state-run enterprise had changed its role from representing "learning from the Soviet Union" to denouncing the Soviet model which was supposed to attach too much importance to the expert.

In the 1970s, we see yet another turn in the role played by the AnSteel and the AnSteel Charter in knowledge production: now they stood for high political consciousness, which was believed to be the driving force of technological innovation and economic production. In *Steel Man Riding Iron Horse on the New Road* (*Gangren tiema kua xincheng* 钢人铁马跨新程, 1974), a volume containing articles introducing new experiences of technological innovation "on the front" of metallurgy, the AnSteel's increased productivity was attributed to the application of the Charter. By 1973, one article claimed, steel, pig iron, and iron ore produced in Anshan had increased by 37 percent or 90 percent in comparison with those produced in 1965. Workers formed shock work (*huizhan* 会战) groups with technicians and managerial members and repaired and restored the facilities—instead of asking for investment from the state—to increase productivity. The article declared that developing political consciousness was a better way than material rewards to initiate workers' activism and creativity.

The return to "shock work," which depends on the extra workload of manpower and which the mass campaigns of technological innovation in 1954 and the revolutions in technology and culture in 1958 intended to do away with, indicated the general failure of the AnSteel Charter in economic production. A 1975 *People's Daily* editorial urged the reader to study theory in order to bring up production.[66] "Theory" here refers not only to Mao Zedong's essays "On Practice" and "On Contradiction," but also his criticism of revisionism—of Liu Shaoqi and Lin Biao as well as of the Soviet Union.[67] One cannot miss the irony here: Mao's texts on empirical epistemology championing "experiment" and "practice" had turned into petrified dogmas themselves.

Uneasy relations among the expert, the Party, and the worker in knowledge production and dissemination symptomized the tensions and problems in China's negotiations through epistemologies of different origins in order to find pragmatic ways of achieving modernization with resource constraints. Despite radical Maoism in the 1970s, the field of veterinary medicine managed to produce knowledge that would sustain after China's re-professionalization of knowledge in 1978, as we argue in the next chapter.

NOTES

1. For an analysis of the CCP's attempt of class leveling by allowing people from peasant and worker family backgrounds more access to (higher) education in the Mao era, see Andreas, *Rise of the Red Engineers*.

2. Here we use Kaple's brief definition of "High Stalinism" (1945–53): it is a period "when the Soviet Communist Party was completely in control of all spheres of production, as well as all aspects of life outside the factory." Kaple, *Dream of a Red Factory*, 6.

3. Kaple, *Dream of a Red Factory*, 59.

4. Chaliaofu 察廖夫 (G. Tsarev), *Gongchang dang zuzhi zenyang lingdao shehui zhuyi jingsai* 工厂党组织怎样领导社会主义竞赛 (How does the party in factory lead socialist competitions?), trans. Chen Dawei 陈大维 (Beijing: Shidai chubanshe, 1951), 6. For similar arguments, also see other books in the series: Kelimannuofu 克里曼诺夫 (A. Klimanov), *Qiye zhong de dang gongzuo* 企业中的党工作 (The party's work in enterprises), trans. Cao Ying 草婴 (Shanghai: Shidai chubanshe, 1950); Gaojinnawa 高金娜娃 (D. Goginawa), *Qiye zhong de dang guanli* 企业中的党管理 (The party's administration in enterprises), trans. Cao Ying 草婴 (Shanghai: Shidai chubanshe, 1950); Shiqipansiji 施契潘斯基 (W. Shepanskii), *Gongchang zhong qunzhong zhengzhi gongzuo* 工厂中群众政治工作 (The masses' political work in factories), trans. Lin Xiu 林秀 (Shanghai: Shidai chubanshe, 1950).

5. Hiroaki Kuromiya, "Edinonachalie and the Soviet Industrial Manager, 1928–1937," *Soviet Studies* 36, no. 2 (1984): 185–204. Here 185–87.

6. Kuromiya wonders whether the failure was due to the systematic constraints in the Soviet political and industrial structures and whether the party only needed a pretext to remove managers, see Kuromiya, "Edinonachalie and the Soviet Industrial Manager," 198.

7. For an in-depth study of the Stakhanovite movement, see Lewis H. Siegelbaum, *Stakhanovism and the Politics of Productivity in the USSR, 1935–1941* (New York: Cambridge University Press, 1988).

8. Zhang Dai 张岱, *Zenyang zai gong kuang qiye zhong shixing yizhangzhi* 怎样在工矿企业中实行一长制 (How to implement the one-man-management system in industry and mining?) (Beijing: Gongren chubanshe, 1954), 9.

9. William Brugger, *Democracy and Organisation in the Chinese Industrial Enterprise (1948–1953)* (Cambridge: Cambridge University Press, 1976), 188–90.

10. Lewis Siegelbaum, "1929: Shock Workers," *Seventeen Moments in Soviet History*, accessed November 27, 2020, http://soviethistory.msu.edu/1929-2/shock-workers.

11. "Aiguozhuyi shengchan jingsai 爱国主义生产竞赛 (Patriotic production competition)," *Renmin Huabao*, no. 2 (1951): page not indicated. See also Zhonghua quanguo zonggonghui shengchanbu 中华全国总工会生产部, *Shengchan jingsai wenti jianghua* 生产竞赛问题讲话 (Speeches on the Problems of Production Competition) (Beijing: Gongren chubanshe, 1950); Liu Zijiu 刘子久, *Laodong jingsai jianghua* 劳动竞赛讲话 (Speeches on Labor Competition) (Beijing: Gongren chubanshe, 1954).

12. J. V. Stalin, "Emulation and Labour Enthusiasm of the Masses—Forward to E. Mikulina's Pamphlet 'Emulation of the Masses,'" *Pravda*, May 22, 1929. Translation from *J. V. Stalin Works* (Vol. 12, April 1929–June 1930): 114–17. For the Chinese translation of Soviet socialist emulation, see Liening 列宁 (Lenin) and Sidalin 斯大林 (Stalin), *Lun shehuizhuyi jingsai* 论社会主义竞赛 (On socialist emulation), trans. He Shu 合树 et al. (Beijing: Gongren chubanshe, 1955) (Soviet original 1941).

13. Chen Sihe 陈思和, "Ruhe dangjia? Zenyang zuozhu? Chongdu Lu Mei zhibi de huaju 'Hongqi ge' 如何当家？怎样做主？重读鲁煤执笔的话剧《红旗歌》 (How to Become the Master? Rereading the Spoken Drama *Red Flag Song* Written by Lu Mei)," *Zhongguo xiandai wenxue yanjiu congkan* 中国现代文学研究丛刊, no. 4 (2011): 31–42; Qian Ying, "The Shopfloor as Stage: Production Competition, Democracy, and the Unfulfilled Promise of Red Flag Song," *China Perspectives*, no. 2 (2015): 7–14. The play was adapted into an eponymous film in 1950 (directed by Wu Zuguang 吴祖光, 1917–2003).

14. Kaple, *Dream of a Red Factory*, 64–65.

15. For the Chinese-language introduction of the Stakhanovite Movement, see Alekseĭ Grigor'evich Stakhanov (А. Г. Стахáнов 斯達哈諾夫), *Sidahanuofu yundong* 斯达哈诺夫运动 (The Stakhanovite Movement), trans. Sun Siming 孙斯鸣 (Shanghai: Shanghai shidai shuju, 1949). A similar understanding of the Stakhanovite Movement appeared also in the *History of the All-Union Communist Party (Bolsheviks): Short Course*, originally published in 1938.

16. Wei Wei 魏巍, "Zou zai shijian de qianmian—gei gongren Wang Chonglun tongzhi de xin 走在时间的前面—给工人王崇伦同志的信 (Walking ahead of time—A letter to the worker comrade Wang Chonglun)," *Renmin Ribao*, December 23, 1953, p. 3.

17. Wang Chonglun 王崇伦, "Wo yao jianjue shixian ziji de baozheng jihua 我要坚决实现自己的保证计划 (I guarantee to realize my production plan)," *Zhonggongye tongxun*, no. 13 (1954): 38–39.

18. Wang Chonglun and Xu Binzhang 许彬章, *Rang women he shijian saipao* 让我们和时间赛跑 (Let's compete with time) (Beijing: Gongren chubanshe, 1954).

19. "Wang Chonglun chuangzao wanyong gongjutai 王崇伦创造万用工具胎 (Wang Chonglun creates the universal fixture)," *Jixie gongren*, no. 1 (1954): 7–9.

20. "Yinian neng wancheng sannian shengchan renwu de Wang Chonglun 一年能完成三年生产任务的王崇伦 (Wang Chonglun, who is able to complete the production quota of three years in one year)," *Jixie gongren*, no. 1 (1954): 5–6.

21. "Anshan Gangtie zongchang kaiban Wang Chonglun xianjin shengchanzhe xuexiao, jinyibu tuiguang Wang Chonglun shengchan gexin jingshen he jingyan 鞍山钢铁总厂开办王崇伦先进生产者学校，进一步推广王崇伦生产革新精神和经验 (The AnSteel Main Plant opens Wang Chonglun advanced producer school to further promote Wang's spirit and experiences of technological innovation)," *Renmin Ribao*, February 8, 1954.

22. Yang Qingchun 杨庆春, "Women shi zenyang tuiguang Wang Chonglun de xianjin jingyan de 我们是怎样推广王崇伦的先进经验的 (How do we popularize Wang Chonglun's advanced experience?)," *Renmin Ribao*, February 8, 1954.

23. "Changsha shi ge chang xuanchuan xuexi Wang Chonglun gexin jingshen de qingkuang 长沙市各厂宣传学习王崇伦革新精神的情况" (Situations of factories promoting and learning Wang Chonglun's spirit of innovation in the city of Changsha)," *Neibu cankao*, April 20, 1954, 207–8.

24. "Fayang Wang Chonglun de gongzuo jingshen, tiqian wancheng guojia jihua 发扬王崇伦的工作精神，提前完成国家计划 (Carry on Wang Chonglun's spirit, complete the national plan ahead of schedule)," *Renmin Ribao*, February 8, 1954.

25. Editorials, "Xiang jishu geming jinjun 向技术革命进军," *Renmin Ribao*, June 3, 1958; "Wenhua geming kaishi le 文化革命开始了," *Renmin Ribao*, June 9, 1958.

26. Cf. Andreas, *Rise of the Red Engineers*.

27. *Gongye de jichu—Gangtie gongye* 工业的基础 —钢铁工业 (Foundation of industry—the iron and steel industry) (Beijing: Zhonghua quanguan kexue jishu puji chubanshe, 1954), 1. It is one of the pamphlets on the industries of steel and iron in the series of Scientific Knowledge of Socialist Industrialization (*Shehuizhuyi gongyehua kexue zhishi* 社会主义工业化科学知识) distributed by All China Association of Science and Technology Dissemination.

28. "Keith D. Millis: The Father of Ductile Iron," *The Free Library*, October 1, 1998, accessed January 20, 2021, https://www.thefreelibrary.com/Keith D. Millis: the father of ductile iron.-a021265468. Chinese scholars contested that similar technology could be found in iron artifacts made in late Western Han Dynasty (c. 202 BC–c. 8 AD), see Hao Shijian 郝石坚, "Qianyan 前言 (Preface)," in *Xiandai qiumo zhutie* 现代球墨铸铁 (Modern nodular cast iron), ed. Hao Shijian (Beijing: Meitan gongye chubanshe, 1989), no page indicated.

29. "Qiumo zhutie 球墨铸铁 (Nodular cast iron)," *Renmin Huabao*, no. 3 (1952).

30. "Yunshushi shang yi chuangju qiumo zhutie zuo ganggui 运输史上一创举 球墨铸铁做钢轨 (A pioneering work in the history of transportation: nodular cast iron used for steel rail lines)," *Renmin Ribao*, September 5, 1958.

31. Wang Daohan 汪道涵, "Zai jiqi zhizao gongye zhong dali tuiguang qiumo zhutie 在机器制造工业中大力推广球墨铸铁 (Vigorously popularize nodular cast iron in the machine building industry)," *Zhishi jiushi liliang*, no. 12 (1958): 1–2; Yu Guangsheng 余光生, "Tiegui yu ganggui bingju, jiasu tielu jianshe 铁轨与钢轨并举，加速铁路建设 (Simultaneously developing railroad tracks and steel rails, accelerating railroad construction)," *Zhishi jiushi liliang*, no. 12 (1958): 3. Wang

Daohan studied mechanical engineering and mathematics in Shanghai before he went to Yan'an in 1937. In the PRC, he worked mainly in the field of trade and investment. In 1981, he was appointed the mayor of Shanghai. Yu Guangsheng studied civil engineering at Shanghai Jiaotong University and obtained his master in rail transportation in 1930 in Michigan. He joined the Communist Party in the United States in 1932 and went to Yan'an in 1940, where he served as an editor of the *Liberation Daily*.

32. Qinghua daxue zhugong jiaoyanzu 清华大学铸工教研组, "Yizhong xinxing jiegou cailiao—qiumo zhutie 一种新型结构材料－球墨铸铁 (A new type of structural material—Nodular cast iron)," 4–5; "Qiumo zhutie de qiuhua chuli 球墨铸铁的球化处理 (Nodularizing treatment of nodular cast iron)," 10–11; "Xin qiuhua ji－xitiemei hejin 新球化剂－矽铁镁合金 (New nodularizing element—Silicon-Iron-Magnesium-Alloy)," 10–11, "Qiumo zhutie tiegui 球墨铸铁铁轨 (Nodular cast iron railroad tracks)," 6, and "Zai yali xia jia mei chuli qiumo zhutie 在压力下加镁处理球墨铸铁 (Processing nodular cast iron with magnesium under pressure)," 8–9, in *Zhishi jiushi liliang*, no. 12 (1958).

33. Diyi jixie gongyebu di wu shejiyuan tiegui yanjiu xiaozu 第一机械工业部第五设计院铁轨研究小组, "Yong shengtie jiaozhu tiegui 用生铁浇注铁轨 (Using pig iron in casting railroad tracks)," *Zhishi jiushi liliang*, no. 12 (1958): 7; Yang Changzu 杨昌组, "Tutie dai gang 土铁代钢 (Native iron replaces steel)," *Zhishi jiushi liliang*, no. 12 (1958): 12.

34. Wang Daohan, "Zai jiqi zhizao gongye zhong dali tuiguang qiumo zhutie," 1.

35. Diyi jixie gongyebu di wu shejiyuan tiegui yanjiu xiaozu, "Yong shengtie jiaozhu tiegui."

36. Cover page, *Zhishi jiushi liliang* 知识就是力量 (*Knowledge Is Power*), no. 12 (1958). On "combining education and productive labor" in Qinghua University in the Great Leap Forward years, see Andreas, *Rise of the Red Engineers*, 52–54.

37. See, for example, Shi Bainian 史柏年, "1958 nian dalian gangtie yundong shuping 1958 年大炼钢铁运动述评 (A critical review of the 1958 Great Leap Forward in Iron and Steel)," *Zhongguo jingjishi yanjiu* 2 (1990): 124–33; and Luo Pinghan 罗平汉, "1958 nian quanmin daliangang 一九五八年全民大炼钢 (A national campaign of Great Leap Forward in steelmaking in 1958)," *Dangshi wenyuan* 11 (2014): 24–31.

38. Ghosh, *Making it Count*, 12.

39. M. Gardner Clark, *Development of China's Steel Industry and Soviet Technical Aid* (Ithaca: Cornell University Press, 1973), 70. Quoted by Chan, *Mao's Crusade*, 185.

40. "Tutie nenggou liancheng gang ye bixu liancheng gang 土铁能够炼成钢也必须炼成钢 (Native iron is capable of making steel and has to make steel)," *Yejinbao*, no. 39 (1958): 3–11.

41. "Cujin tutie lian haogang 促进土铁炼好钢 (Promoting native iron to make steel)," *Yejinbao*, no. 39 (1958): 12; "Shanghai Tianjin gangtie gongren po lian'gang nanguan, tutie nenggou lianchu haogang 上海天津钢铁工人大破炼钢难关，土铁能够炼出好钢 (Shanghai and Tianjin steel workers get breakthrough in steelmaking, native iron is capable of making good steel)," *Renmin Ribao*, October 3, 1958.

42. Wu Guangya 吴光亚, "Tutie liangang de guanjian wenti 土铁炼钢的关键问题 (Crucial issues of native iron making steel)," *Kexue huabao*, no. 11 (1958): 401–2, 423.

43. "Xinxing gangchang zai po jishuguan, tutie liangang hegelü 95% 新兴钢厂再破技术关, 土铁炼钢合格率95% (Xinxing Steelworks made a breakthrough in technology again, 95% of the steel made from native iron reached the standard)," *Yejinbao*, no. 4 (1959): 21.

44. "Tutie tulu lian haogang 土铁土炉炼好钢 (Native iron and native furnace make good steel)," *Jixie gongren*, no. 1 (1959): 23.

45. "Tutie tugang ji neng chengcai you neng chengqi 土铁土钢既能成才又能成器 (Native iron and native steel are good and useful materials)," *Yejinbao*, no. 3 (1959): 31–34.

46. Cermet, a composite material made by mixing, pressing, and sintering metal with ceramic, was first developed in the United States after WWII.

47. Han Qilou 韩棋楼, "Taoci xian shentong, neng ding gangtie yong 陶瓷显神通, 能顶钢铁用 (Ceramics reveal remarkable characteristics, can be used to substitute steel)," *Kexue huabao* 11 (1958): 417.

48. Zhang Chaozong 张朝宗, "Taoci nianqing le 陶瓷年轻了 (Ceramic is becoming young)," *Zhishi jiushi liliang*, no. 4 (1959): 1–2.

49. "Yi ci dai gang dazao jixie shebei 以瓷代钢大造机械设备 (Using ceramic to substitute steel in building mechanical equipment)," *Renmin Ribao*, February 15, 1959, 6.

50. Wang Tianbao 王天宝, "Bai gang 白钢 (White Steel)," *Kexue Huabao*, no. 10 (1962): 388–91.

51. Zhang Heping 张和平, "Yi ci dai gang de taoci tuogun wenshi 以瓷代钢的陶瓷托辊问世 (The ceramic roller came to the world)," *Renmin Ribao*, January 25, 1992, 2

52. Han Qilou, "Taoci xian shentong, neng ding gangtie yong," 417; and the spread titled "Yi tao dai gang 以陶代钢 (Replacing steel with ceramic)" in the same issue.

53. Matsumoto, Toshiro, "Continuity and Discontinuity from the 1930s to the 1950s in Northeast China: The 'Miraculous' Rehabilitation of the Anshan Iron and Steel Company Immediately After the Chinese Civil War," in *The International Order of Asia in the 1930s and 1950s*, eds. Shigeru Akita and Nicholas J. White (Surrey: Ashgate, 2010), 255–73.

54. For example, see Yang Wenzhong 杨文仲, "Xin Zhongguo de zhonggongye 新中国的重工业 (Heavy industry of the new China)," in *Xin Zhongguo huifu shiqi de zhonggongye jianshe* 新中国恢复时期的重工业建设 (Heavy industry construction in the recovery period of the new China), ed. Yang Wenzhong (Beijing: Sanlian shudian, 1954), 1–15.

55. Gao Hua 高華, "Angang xianfa de lishi zhenshi yu 'zhengzhi zhengquexing' 鞍鋼憲法的歷史真實與「政治正確性」," *Ershiyi shiji*, no. 58 (2000): 62–69; William A. Byrd, *Chinese Industrial Firms under Reform* (Oxford: Oxford University Press, 1992), 306–7.

56. "Weile guojia gongyehua, kaizhan jishu gexin yundong—Qingzhu Angang jishu gexin zhanlanhui kaimu 为了国家工业化，开展技术革新运动－庆祝鞍钢技术革新展览会开幕 (Launching the technological innovation campaign for national industrialization—Celebrating the opening of the AnSteel's technological innovation exhibition)," *Renmin Ribao*, April 16, 1954.

57. See, for example, the collection *Xuexi Angang jishu gexin jingyan* 学习鞍钢技术革新经验 (Studying AnSteel's experiences in technological innovation) (Chongqing: Chongqing renmin chubanshe, 1954). It contains twelve articles by the visitors of the Exhibition of Angang Technological Innovations who elaborated on their understandings of technological innovations.

58. Ma Chengde 马成德, "Xuexi Sulian, kaizhan jishu gexin yundong 学习苏联，开展技术革新运动 (Studying the Soviet Union, launching the technological innovation campaign)," *Zhonggongye tongxun*, no. 13 (1954): 16–20.

59. Anshan gangtie gongsi 鞍山钢铁公司, "Anshan gangtie gongsi de jishu gexin yundong 鞍山钢铁公司的技术革新运动 (The campaign of technological innovations in the Anshan Iron and Steel Works)," *Zhonggongye tongxun*, no. 13 (1954): 11–15.

60. Zhang Tianlai 张天来 et al., *Zai kexue jinjun de daolu shang* 在科学进军的道路上 (On the Road of Marching towards Science) (Shenyang: Liaoning renmin chubanshe, 1956).

61. Lai Denong 赖德浓, "Sulian zhuanjia dui Angang yeyu gongye daxue de yijian 苏联专家对鞍钢业余工业大学的意见 (Soviet experts advised on the spare-time industrial university of the Ansteel)," *Neibu cankao*, February 28, 1955, 273–74.

62. Gao Hua, "Angang xianfa de lishi zhenshi yu 'zhengzhi zhengquexing,'" especially 65.

63. The Party committee of the municipal government of Anshan had been responsible for the Party organization work in the AnSteel except for the short time between 1954 and 1955, when the AnSteel briefly established its own Party committee within the company. See Li Huazhong 李华忠 and Zhang Yu 张羽, *Angang sishi nian* 鞍钢四十年 (Forty Years of the AnSteel) (Shenyang: Liaoning renmin chubanshe, 1989), 40.

64. For a complete quote of Mao's written instruction, see Dai Maolin 戴茂林, "Angang xianfa yanjiu 鞍钢宪法研究," *Zhonggong dangshi yanjiu*, no. 6 (1999): 38–43.

65. See, for example, Stephen Kotkin, *Magnetic Mountain: Stalinism as a Civilization* (Berkeley: University of California Press, 1995); Oleg V. Khlevniuk, *In Stalin's Shadow: The Career of "Sergo" Ordzhonikidze* (Armonk: M. E. Sharpe, 1995); Siegelbaum, *Stakhanovism and the Politics of Productivity in the USSR.*

66. "Zhua lilun xuexi, cu gongye shengchan 抓理论学习，促工业生产 (Studying theories to promote industrial production)," *Renmin Ribao*, March 11, 1975, collected in Yejin gongye chubanshe 冶金工业出版社, *Zai Angang xianfa de guanghui qizhi xia qianjin* 在鞍钢宪法的光辉旗帜下前进 (Marching under the glorious flag of the AnSteel Charter) (Beijing: Yejin gongye chubanshe, 1975), 1–3.

67. The listed revisionist opinions included: material incentives that subverted the solidarity among workers; the authority of the specialist that privileged mental labor over manual labor; and the capitalist-styled pursuit of profit. Also see *Zai Angang xianfa de guanghui qizhi xia qianjin*, 1975

Chapter Five

Creating a Bifurcated Knowledge System
The Case of Chinese Veterinary Medicine

In 1950, an article published in the journal *Agriculture in the Northeast* (*Dongbei Nongye* 东北农业) described the difficulties faced by the newly founded Association for Veterinary Medicine (*Shouyi lianhehui* 兽医联合会) in educating veterinary practitioners. Arguing that the problem lay less in developing political consciousness than disseminating veterinary knowledge, the author asked how should the students be taught to integrate knowledges that did not necessarily conform to the principles of Western veterinary medicine ordained by the state in the past?[1]

As we have discussed in the preceding two chapters, the CCP's abandonment of the Soviet model of developing science and technology as well as its advocacy and implementation of "mass science" during the late 1950s and 1960s attempted to realize a widespread transformation in knowledge epistemology that valorizes non-expert knowledge presumably created by the masses. Different from the cases of tool reform and steel production innovation, where productivity was increased by incremental technological improvements, veterinary medicine—similar to human medicine—encountered a fundamental problem of justifying and reconciling knowledges of different epistemological origins. Since the Republican era, modern biomedicine was set in stark opposition to so-called superstitious medical practices, as Ralph Croizier argued in his 1968 study *Traditional Medicine in Modern China: Science, Nationalism, and the Tensions of Cultural Change*, a work that details the efforts of the medical elites to abolish Chinese medicine. This changed in 1955 with the founding of the Academy of Chinese Medical Sciences (*Zhongyi yanjiuyuan*), the first state-founded institution of Chinese medicine. It was a clear signal to follow the call of Mao Zedong to combine both new and old medicine, Chinese and Western medicine to build a united front of medical workers for the socialist construction of the country. This front was intended to become a low-cost solution to health care in the countryside, relying primarily on indigenous herbs and acupuncture.[2] Chinese medicine became fully institutionalized and standardized by 1963 and was practiced and taught in hospitals and schools.[3] The barefoot doctors came to play an important role in the dissemination and application of Chinese medical knowledge. According to Fang Xiaoping the barefoot doctor program—though said to have collapsed with the market reforms and commercialization of health care after 1978[4]—contributed to the arrival of Western medicine to villages that had long been dominated by Chinese medicine.[5] Fang's approach to the barefoot doctor practice assumes two discernible schools of medical knowledge that are more or less systematic in their own ways and can thus be differentiated. Volker Scheid,[6] on the contrary, has identified in his ethnographical account a medical pluralism in human medicine that we can also find in veterinary medicine. In this pluralism, we are not trying to construct Chinese veterinary medicine (CVM) as the "other" form of veterinary medicine, but rather focus on identifying the various actors involved in the knowledge production and dissemination of it while highlighting the establishment of a bifurcated field of veterinary medicine in the Mao era.

This chapter asks how the state determined the relation between Western and indigenous veterinary knowledge that barefoot doctors for veterinary medicine (*shouyi chijiao yisheng* 兽医赤脚医生) were supposed to disseminate in the countryside. We first delineate the institutionalization of a knowledge system through administrative and academic measures in the 1950s and 1960s, showing the state's intervention with integrating CVM into the discipline of veterinary biomedicine in the PRC. Then we look at the dissemination of veterinary knowledge through various forms, including discussions in manuals and among practitioners. As an example, we offer an analysis of two publications in the 1970s dealing with swine diseases to show specific ways how Western and Chinese veterinary medicines were related to each other. We argue that the success of CVM was the result of the epistemological turn that preferred practice over theory and—quite different from the case of human medicine—bypassed the binaries of science/superstition and foreign/indigenous knowledge.

INSTITUTIONALIZING VETERINARY KNOWLEDGE IN THE MAO ERA

After the founding of the PRC the party-state quickly realized that epidemics and diseases were posing a considerable risk to animal husbandry. Mao Zedong had already declared in 1942: "The greatest enemies of animal husbandry are too many diseases and too little fodder. Its development cannot be realized if these two problems are not solved."[7] Accordingly, the treatment of veterinary diseases stood high on the state's agendas during the 1950s. Contagious diseases such as hog cholera (*zhuwen* 猪瘟) not only caused huge economic losses, but was also seen as a potential danger to human health.[8] Thus, the party-state decided to intensify the training of veterinary personnel.

Early discussions on the reform of veterinary medicine saw it imperative to combine both Western and Chinese veterinary medicine, as stated by a contribution in the *People's Daily* in July 1950. This short article reported on the situation of training veterinarians. At a conference at Beijing Agriculture University (*Beijing Nongye Daxue* 北京农业大学), leaders and teachers of thirty-five veterinary working stations in and around Beijing met to "sum up work experiences" (*zongjie gongzuo jingyan* 总结工作经验) and to discuss how Chinese and Western veterinary medicine practitioners should combine their experiences. Yue Tianyu 乐天宇 (1901–84), a biologist and dean of the university, told his audience while reporting on the Communists' experience of veterinary medicine education that Western veterinarians were first hired to help cure sick animals for farmers but were not overtly effective. In his view, the combination of Western and Chinese veterinary knowledge was a pragmatic option to deal with the shortage of personnel and medicine. The decision to include Chinese veterinarians in the knowledge dissemination movement was apparently supported and welcomed by rural residents.[9]

A landmark directive was issued by premier Zhou Enlai in the name of the State Council on January 5, 1956. The "State Council's Directive of Strengthening the Work of Folk Veterinarians" (*Guowuyuan guanyu jiaqiang minjian shouyi gongzuo de zhishi* 国务院关于加强民间兽医工作的指示) declared the goal of largely eradicating and controlling major contagious disease of domestic animals in the coming seven years. The Ministries of Agriculture and Public Heath were made responsible for mobilizing folk veterinarians when organizing agricultural production. The reason was again the continuous shortage of "new veterinarians," namely only several thousands in contrast to more than 80,000 "folk veterinarians" in the countryside who were practicing half-agriculture and half-medicine, Zhou pointed out. The directive demanded that folk veterinary medicine practitioners (*minjian shouyi renyuan* 民间兽医人员) should be organized to improve their professional knowledge and political consciousness of serving the people. Their teaching should include medical theory and new techniques such as animal vaccination. CVM research should be carried out systematically by using new scientific theory and methods in order to study and summarize the techniques and experiences (*jingyan*) of veterinarians.[10] In addition, tested and effective formulae should be promoted and disseminated; and the research process should involve renowned folk veterinarians who should be mobilized to train apprentices and offered remuneration for their service to society.[11]

It is notable in this directive that new terms are used: "new veterinary medicine" (*xin de shouyi kexue* 新的兽医科学) and "folk veterinary medicine" (*minjian shouyi* 民间兽医) replaced the categories of Western and Chinese medicines (西医，中医), which in the case of human medicine had been interpreted along the lines of science vs. superstition, as we have shown with the lantern slideshow material "Preventing Contagious Diseases in the Summer Season" in chapter 1 (figures 1.1–1.4). By doing so, the directive not only downplayed the Western/Chinese binary, but also related Chinese veterinary practitioners to the political notion of the masses. More important, the directive suggested ways of combining the two veterinary medicines in terms of practice, research, and knowledge transmission. Whereas it called for research on Chinese veterinary medicine in the framework of "new scientific theory and methods" and the cooperation between new and folk veterinarians, its acknowledgment of proven effects (*queyou chengxiao* 确有成效) of CVM indicates the official promotion of a bifurcated knowledge system.[12]

A report on the folk veterinarian Wang Daofu 王道福 in early 1957 in *People's Daily* fleshes out the ideal of a folk veterinarian's transformation. Wang's pride in thousands of years of medical practice was combined with a professional modesty. Admitting that Chinese veterinary practices lacked some scientific procedures such as measuring body temperature and disinfection during castration, Wang repeated the official position that folk and Western veterinary medicines should learn from each other. He joined the organization of folk veterinary medicines, shared his family recipes, and worked with other folk veterinarians to sort out more than sixty effective recipes.[13]

In 1957, Xiong Dashi 熊大仕 (T. S. Hsiung, 1900–87), professor in veterinary medicine at Beijing Agriculture University, who received his medical degree in Veterinary Medicine at Iowa State University in 1927, and editor of the influential *Newsletter of Folk Veterinary Medicine* (*Minjian shouyi tongxun* 民间兽医通讯),[14] proposed to integrate Chinese veterinary knowledge into the discipline of veterinary science. He suggested that a department be set up in the newly founded Chinese Academy of Agricultural Studies to lead the research in CVM nation-wide; he asked for institutional support from the central government to facilitate hiring experienced folk veterinarians, publishing veterinary classics, and setting up research and teaching groups in veterinary medicine high schools to design a curriculum for veterinary training. Xiong, furthermore, reiterated the need of improving the social and financial status of Chinese veterinarians, which implies that these points mentioned in the 1956 directive may have not been carried out properly. If these problems were neglected, there would be fewer and fewer practitioners, he warned.¹⁵ Regarding the combinations of Chinese and Western veterinary medicines as well as teaching and research as key issues to enrich Chinese modern veterinary science (*woguo xiandai shouyi kexue* 我国现代兽医科学), Xiong proposed concrete steps to institutionalize Chinese veterinary medicine by acknowledging the achievements of the past.

This included the publication of historical sources and ancient veterinary treatises. In 1957, the *Yuan-Heng Therapeutic Treatise of Horses, with Appended Treatises on Oxen and Camels* (*Yuan-Heng liaoma ji: fu niu tuo jing* 元亨疗马集: 附牛驼经), compiled by the brothers Yu Benyuan 喻本元 and Yu Benheng 喻本亨 in the Ming Dynasty (1368–1644), was republished. This book is believed to be the most widespread treatise on veterinary diseases, presenting a detailed overview on diagnosis, treatment by acupuncture, drugs, and moxibustion. Published in April 1958, the *Collection of Materials on the History of Animal Husbandry in China* (*Zhongguo xumushi ziliao* 中国畜牧史资料) by Wang Yuhu 王毓瑚 (1907–80)¹⁶ uses "knowledge and techniques" (*zhishi he jishu* 知识和技术) to categorize the materials collected on the treatment of domestic animals while shying away from the term "science" (*kexue*).¹⁷ The first section of the book presents a historical overview on animal husbandry from the pre-Qin era to the Qing dynasty (1644–1911); and the second section deals with knowledge and techniques for breeding, nourishing, and medical treatment for the animals such as horse, donkey, mule, cattle, sheep/goat, pig, chicken, goose, and duck. In his preface, Wang traced the idea of compiling the collection to a Soviet expert at Beijing Agriculture University in the second half of 1955. This expert—who is not mentioned by name—had encouraged him to compile a short history of husbandry in China. Wang noted that many prescriptions are transmitted orally by veterinary practitioners in the countryside. Due to the lack of written records, their knowledge and techniques were likely to disappear (a complaint often observed in the prefaces of veterinary texts of the 1950s and 1960s), which urged him to collect unrecorded prescriptions and knowledge from various ethnic groups outside urban centers.¹⁸

A report on the first nationwide conference on veterinary research work held in Lanzhou in 1958 adopted an even more self-confident tone highlighting the importance of Chinese veterinary medicine. The combination of Western veterinary medicine (*xiyang shouyi* 西洋兽医) with its Chinese counterpart, the report stated, meant first of all that the former should learn from the latter and make use of the rich national cultural heritage that supposedly dated back to the Spring-and-Autumn era (771–476 BC). The report then referred to international recognition of CVM in the Soviet Union, Czechoslovakia, Vietnam, North Korea and East Germany (GDR) to champion the validity of Chinese medical practice.¹⁹ In rare cases publications also carried praise from foreign experts, such as in the 1958 booklet *Practical Acupuncture and Moxibustion in Veterinary Medicine* (*Shiyong shouyi zhenjiuxue* 实用兽医针灸学) that included a preface in Russian and Chinese contributed by the Soviet expert A. A. Alenkowitsch.[20]

In 1959, the National Conference on Research in Chinese and Western Veterinary Medicine (*Quanguo Zhong-Xi shouyi yanjiu gongzuo huiyi* 全国中西兽医研究工作会议) under the leadership of Nie Rongzhen foregrounded the guidelines of veterinary medicines as "scientific research serving [economic] production, theory combined with practice, and Chinese veterinary medicine combined with Western one, and specialist research institutions combined with mass scientific activities." These guidelines thus integrated CVM into "science" for its economic function of serving production, epistemological justification of practicality, and political agenda of mobilizing the masses. The professional achievements listed on the conference included not only newly developed vaccinations but also acupuncture and effective medical drugs, all attributed to the improved political consciousness of the practitioners as the result of a series of mass movements in previous years. Among the important research tasks for the year 1959, the conference named "researching on the techniques and theory of Chinese veterinary medicine and pharmacology, collecting and sorting out Chinese veterinary prescriptions and effective formulae, and summarizing the techniques of acupuncture and experiences of diagnosis and treatment in CVM."[21] The major reason of setting up these tasks, according to a note of the Ministry of Agriculture issued on February 5, 1959, was that CVM knowledge had not been systematically preserved in written texts and mostly were transmitted orally from master to disciples,[22] a point also mentioned by Wang Yuhu.

It is for this reason that the State Council issued in 1963 a new directive that addressed the problematic situation of veterinary medicine during and after the Great Leap. It reiterated the problems already recognized in 1956 that the lack of organization and renumeration constituted a great problem for further development of the discipline. Unorganized doctors tended to demand higher fees given the fact that they were working individually and were exposed to considerable economic risks. This made it necessary for the state to intervene and to pursue the policy aiming at uniting veterinary practitioners, emphasizing the practical use of medicine, educating practitioners, and raising the level of veterinary knowledge.[23]

The new directive led to a republication of the *Yuan-Heng Therapeutic Treatise* in 1963. Published under the title *Newly Edited and Corrected Yuan-Heng Therapeutic Treatise of Horses, Oxen and Camels* (Chongbian jiaozheng Yuan-Heng liaoma niu tuo jing quanji) 重编校正元亨疗马牛驼经全集, it was amended with commentaries made by the Institute of Chinese

Veterinary Medicine at the Chinese Academy of Agricultural Sciences. Its editorial preface warned the reader not to simply imitate practices of the past described here, but rather detect the "feudal superstitions contradicting science" (*weifan kexue de fengjian mixin sixiang* 违反科学的封建迷信思想),[24] such as the statements that different coloring on the forehead or a white spot on the chest of a cow were a good omen, etc.[25] Except for this short warning, the remainder of this book does not use science (or technology) as a relevant epistemic criterion. On the contrary, the *Yuan-Heng Therapeutic Treatise*—as many others—is rather seen as a valuable contribution representing the "veterinary medicine of our ancestral land" (祖国的兽医学). To republish this and other books meant for them mirroring the farmers' experiences of the past.[26]

Nathan Sivin has argued in his study on *Traditional Medicine in Contemporary China* (1987) that the knowledge in books on traditional medicine presents itself as authoritative because it is considered timeless. The publication of ancient veterinary medicinal works turns experiences and practices of the past into timeless, de-historicized knowledge. It is thus able to translate this knowledge into the present and dispose of it on the basis of daily needs, without having to worry about or being restrained by epistemological concerns. This became the central characteristic of the majority of veterinary dissemination materials in the Mao era as we show in the following.

DISSEMINATING VETERINARY MEDICINE

Manuals and handbooks dealing with problems in animal husbandry, ranging from preventing and treating animal diseases to increasing the reproduction rate by artificial insemination, mushroomed in the decades after 1949. Written in simple language and often illustrated with anatomical drawings, these texts played a central role in disseminating veterinary knowledge in the rural areas, where they were often combined with exhibitions and research reports by radio and in newspapers.[27]

A classic example of such books is the *Tested Drug Formulae of Folk Veterinary Medicine* (*Minjian shouyi yanfang* 民间兽医验方), published in 1958 by the Department of Agriculture in Heilongjiang Province, which seemed popular enough to be followed by a second collection in 1959. Though the preface is deeply immersed in the rhetoric of class struggle, the book itself makes as its aim to collect therapeutic methods, testing them and then finally introducing the promising methods to its readership. The short and straightforward introduction—included by political necessity—is followed by a long list of cures arranged according to the types of disease. In most cases the reader is advised to collect and grind medical herbs, infuse them in hot water, and make animals drink the potion. Sporadically, the names of the peasant or the veterinarian who contributed the formula are mentioned,[28] apparently to stress their mass-line, grassroots feature. There are no explanations whatsoever about the origin of the drug formula or any scientific analysis of the ingredients, and even the fact that the different diseases are organized according to the taxonomy of Western medicine (diseases of the digestive/nervous/respiratory system) is not problematized.[29]

A 1958 booklet on veterinary drug formulas from Anhui emphasized that the recipes had been collected from animal husbandry workers during a province-level conference in 1956 (*Quansheng xumu shouyi gongzuozhe daibiaohui* 全省畜牧兽医工作者代表会). It was published after these recipes underwent scientific verification procedures (*women dui zhexie danfang you jinxingle kexue lunzheng* 我们对这些单方又进行了科学论证) and appeared first in the form of internal publication. It was distributed to animal husbandry workers who were asked to test the formulae and submit their assessment back to the conference, which on a second meeting deliberated about the feedback before distributing the booklet officially.[30]

Such manuals are often supplemented by field reports (*jingyan baogao* 经验报告) that reproduce experiences of farmers in dealing with specific problems, thereby again reinforcing the idea that knowledge was derived from the practical experiences of the working masses.[31] The field reports aim at convincing the readers by pointing out that the new knowledge—not necessarily produced by experts in the ivory tower—was functional even though some of the new methods were highly unorthodox, like feeding pigs with cow excrement.[32]

Such an example is by no means exceptional. The mass-line policy made it imperative to apply knowledge from local practical experiences[33] if it was to become a functional alternative to biomedicine. Neither alternative was exclusive, as shows the long list of prescriptions in the books that were—contrary to the disease taxonomy—not categorized along the lines of Western or Chinese veterinary medicine. For instance, lice-stricken pigs are treated with either a 5 or 7 percent solution of DDT (rubbed on the infested spots), by applying a lotion made from cooking tobacco leaves in 90 parts of water, or a lotion made from the CM herb stemona tuberosa (*baibu* 百部) and sorghum wine (*gaoliang jiu* 高粱酒), a 1972 book on *Prevention and Treatment of Swine Diseases* (*Zhubing fangzhi* 猪病防治) claimed.[34] Likewise, a rabitt's bladder infection can be treated with anti-inflammatory tablets or with a mix of a variety of Chinese herbs such as grass-leaf sweet flag (*Acorus gramineus* 石菖蒲), Akebia (*Akebia trifoliate* 白木通), capillary wormwood (*Artemisia capillaris* 茵陈蒿), chrysantheme (*Chrysanthemum morifolium* 杭菊), and Phellodendron tree bark (*Cortex Phellodendri* 黄柏), as listed in a manual on preventing veterinary diseases.³⁵

In some cases, traditional cures even defied results of biomedical diagnosis. A handbook of 1959 insisted that if a disease is caused by a lack of vitamins, the reasons is not to be found in a lack of vitamins in the fodder, but in the very fact that the organism cannot absorb them. It is thus imperative to strengthen the nerves that influence the absorption process which can be achieved by acupuncture and moxibustion.³⁶ A 1957 publication resulting from the 1956 National Conference on Folk Veterinary Medicine presents a formula for treating anthrax in cows. After describing the symptoms and identifying bacteria as the source of disease the booklet details eleven drug formulae, including one from Yunnan Province that proposed feeding the cow with chicken blood, or if not available, the cow's own blood. The undiscriminating enumeration of treatments from different regions of the country was justified by the large degree of regional differences that makes the availability of certain herbs in all parts of the country difficult. Therapeutic choices must therefore be flexible and take local conditions into consideration, the editors argued.³⁷

Throughout the Mao era publications on veterinary medicine present various kinds of therapeutic knowledge indiscriminately as viable treatments, which also explains why the term *kexue*—contrary to the publications on agricultural mechanization and general works on animal husbandry—appeared much less prominent, except in cases where CVM is praised as part of China's rich heritage of natural sciences (*Woguo yongyou fengfu de ziran kexue lishi yichan* 我国拥有丰富的自然科学历史遗产).³⁸

During the radical phase of the Cultural Revolution the discourse on local practices became intertwined with politics, that is identifying the class enemy who defended a wrong epistemology of knowledge. Liu Shaoqi and Lin Biao, for example, were criticized for their conviction that specialists should take the lead. As the editors of *Prevention and Treatment of Swine Diseases* pointed out, their idealism (*weixinlun* 唯心论) and apriorism (*xianyanlun* 先验论) would only lead to the restoration of capitalism. Instead of combating CVM and pursuing national nihilism (*minzu xuwuzhuyi* 民族虚无主义), indigenous and foreign methods should be combined (*tuyang jiehe* 土洋结合).³⁹ A book titled *Prevention and Treatment of Poultry Diseases* (*Qinbing fangzhi* 禽病防治) dating from 1975 urged its readers not to follow their revisionist line, which preferred foreign to indigenous techniques (*zhongyang qingtu* 重洋轻土).⁴⁰ Though highly rhetorical in language these statements do not worry about the question whether CVM is scientific or not. The handbooks *Prevention and Treatment of Swine Diseases* (1972) and *Preventing and Treating Poultry Diseases* (1975) describe in their prefaces that after finalizing the first draft of the book, inspection teams of the institutions participating in the book compilations were sent to the countryside in Jiangxi and Zhejiang to inquire about the experiences of local farmers and to add information on herbal medication. By accounting for the experiences and knowledge of lower strata of society, so the editors argued, the "revisionist" line of Lin Biao and Liu Shaoqi was refuted while the Maoist epistemological principle of practice and the political agenda of self-reliance were upheld.

The insistence on the mass-line policy was an important factor for the emergence of so-called "barefoot veterinarians" (*chijiao shouyi* 赤脚兽医) in the 1960s. The term—modeled on the barefoot doctor in human medicine (*chijiao yisheng* 赤脚医生)[41]—appeared probably for the first time in a *People's Daily* article in 1968. It reported on an agricultural school in Jiangsu Province that educated children of poor peasants and urban youngsters who had been sent to the countryside. Founded in 1965 the school supplied the reportedly increasing agricultural production with trained agricultural technicians and barefoot veterinarians, with the latter receiving two years of training before joining a work team.[42] Another newspaper article published on May 16, 1969, explained that the barefoot veterinarians trained in communes and veterinary stations came from the same social strata as the middle and poor peasants,[43] and were thus part of the masses sharing the preferred knowledge. A later article published at the end of the Cultural Revolution summarized that the peasants of each brigade had the responsibility for veterinary medicine. Both are said to have a high level of class consciousness and special technical knowledge. It placed great emphasis on the use of Chinese medicinal herbs because such treatments were deemed to combine the benefits of low cost and easy treatment and efficacy.[44] The training was, according to a report in the *People's Daily*, a three-semester program in which the students learned the knowledge about more than twenty animal diseases, treatment of common diseases, and more than fifty acupuncture points.[45]

A discussion of acupuncture in veterinary medicine can be found in the majority of handbooks, even in those textbooks that primarily concentrate on Western medicine.[46] They describe, as many others, potential ailments and their therapy, treating vitamin deficiency syndrome, for example, with either special nutrition, Chinese herbs, or acupuncture that sometimes named over sixty different acupuncture points (*xuewei* 穴位) on the pig's body. Acupuncture in animal husbandry was considered an important contribution of local knowledge, according to the principle of using local materials (*jiudi qucai* 就地取材). It was developed for a number of applications, be they for anesthetization (enhanced by electrical stimulation),[47] treating respiratory system diseases, or contagious diseases.[48] Publications include detailed descriptions of different kind of needles, hints for finding the correct acupuncture points, and suggestions of how to restrain animals during treatment.[49]

In 1960, two veterinarians Fan Yuqi 范玉奇 and Xiong Deyou 熊德有 were credited in an internal (*neibu*) publication with having developed an acupuncture therapy against hog cholera. Having treated more than 3,000 hogs in the past years they are said to have reached a healing rate of 80 percent. The booklet described bacteria as contagious pathogens and introduced diagnostic principles such as body temperature, which are derived from biomedicine, and preventive treatment by vaccination to the farmers,[50] while providing a colored Map of Acupuncture Points of a Pig as shown in figure 5.1.[51]

In general, acupuncture and moxibustion were propagated for three reasons, namely high effectiveness in treatment, simplicity in learning and execution, and cost effectiveness. In the literature on the education of veterinary barefoot doctors, these advantages were constantly mentioned, judged not only as theoretically possible, but politically necessary. In Maoist parlance, barefoot veterinarians were serving the people and were expected to contribute to socialism.[52] Theory and practice were to be connected—by "combining

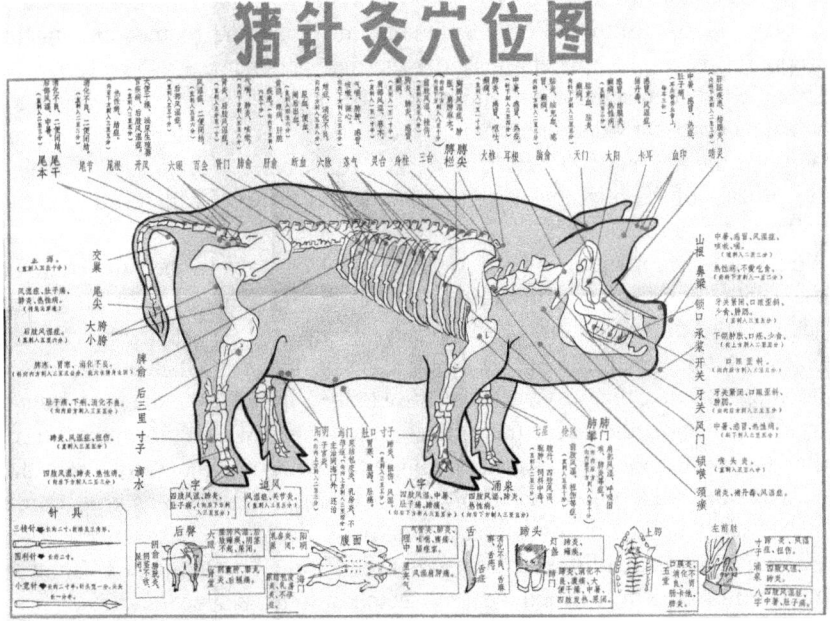

Figure 5.1. Map of Acupuncture Points of a Pig (*Zhu zhenjiu xuewei tu* 猪针灸穴位图, designer unknown), in Jiangsu sheng nongye kexue yanjiusuo geming weiyuanhui 江苏省农业科学研究所革命委员会. *Zhongshouyi zhenliao xuanbian* 中兽医诊疗选编 (Nanjing: Jiangsu sheng geming weiyuanhui, 1970), annex.

the Chinese and Western medicine while mainly using Chinese herbs"[53]—which paved the way for an inclusive version of veterinary medicine that was supposed to end the binary of foreign vs. indigenous medical knowledge.

Taking the 1972 *Prevention and Treatment of Swine Diseases* (*Zhubing fangzhi*) as a case study we can see how the two strains of veterinary knowledge interacted with each other. The book starts with an introduction to the anatomy of the pig, then informs about contagious diseases and hygiene before detailing the four methods of diagnosis (*sizhen* 四诊), that is observation (*wangzhen* 望诊), listening (*wenzhen* 闻诊), inquiring (*wenzhen* 问诊), and palpation (*qiezhen* 切诊). Similar to the case of Chinese medicine defined as a combination of "pattern differentiation and treatment determination" as formulated by Judith Farquhar,[54] diagnosis in CVM is based on an overall analysis of the illness and the patient's condition (*bianzheng shizhi* 辨证施治), as another publication on prevention of pig diseases holds.[55] The remainder of the 1972 book provides treatment plans for various diseases where only CVM treatments are explicitly justified. In the case of acupuncture and Chinese medical herbs (*zhongcaoyao* 中草药) it is the seeming inevitable reference to thousand year long experience, while Western medicine appears to need no justification, implying that its efficacy is taken for granted. Contrary to the opposition of Chinese human medicine and biomedicine in the Republican era, the epistemology of veterinary medicine was not seen as a struggle between science and superstition because it was discussed at a time when its practical value in economic production was central. As a result, not only did the notion of *Zhongshouyi* 中兽医 only emerge after 1949, but also the process of scientizing (*kexuehua*) met less resistance, which is due to the following two reasons.

First, the valorization of experiences occurred only in the latter half of the 1950s, converging with the empowerment of the non-expert under the aegis of the state at a time when veterinary medicine was neither organized nor institutionalized. In contrast to the case of human medicine the institutionalization of Chinese veterinary knowledge did not encounter resistance from an urban medical elite that was educated in the West.[56] Second, its easy diagnostics and cost effectiveness, combined with the labor surplus on the countryside, made time- and labor-intensive procedures such as preparation of herbal medicine, acupuncture, and moxibustion feasible practices. As a consequence, many of the problems identified in the development of Chinese human medicine in twentieth century seem not to have appeared here.

In sum, the PRC state established a bifurcated knowledge system by making CVM a valid knowledge in the larger project of modernization. In addition to government directives, professional scientists worked on building an academic discipline of veterinary medicine integrating CVM through publishing classic literature and academic journals, writing history of its pre-modern "knowledge and techniques," organizing research and training as well as designing curricula. Meanwhile, handbooks and field reports were disseminated and "barefoot" veterinarians were trained for the rural areas. In this process, CVM is justified by the reference to experiences, or *jingyan*, that became part of the veterinary canon in the Mao era precisely because they have an affinity with Mao's concept of practice. This also turned CVM into a site demonstrating the veracity of Marxist class theory, Mao's dialectical materialism, and not the least, the CCP's modernization project. As the result, a Chinese version of veterinary medicine emerged that was able to include medical knowledge derived from different knowledge traditions without seeing them as mutually exclusive. In veterinary work the choice depended on the local situation and the availability of personnel and materials. Such pragmatism further opened up a relatively flexible epistemic space, where knowledge production and dissemination are recognized as a social process that involves more than one actor.

NOTES

1. See here Yu Cong 于匆, "Women shi zheyang tuanjie Zhongshouyi de 我们是这样团结中兽医的 (We have united Chinese veterinary medicine in this way)," *Dongbei nongye*, no. 20 (1950): 24–25.

2. Victor W. Sidel, "The Barefoot Doctors of the People's Republic of China," in: *New England Journal of Medicine* 286, no. 24 (1972): 1292–1300.

3. Taylor, *Chinese Medicine in Early Communist China*, 12, 147.

4. Yang Nianqun, "The Memory of Barefoot Doctor System," in *Governance of Life in Chinese Moral Experience: The Quest of an Adequate Life*, eds. Everett Zhang, Arthur Kleinman, and Tu Weiming (London: Routledge, 2011), 131–45.

5. Fang Xiaoping, *Barefoot Doctors and Western Medicine in China* (Rochester: University of Rochester Press, 2012), 181, 204–5.

6. Scheid, Volker, *Chinese Medicine in Contemporary China: Plurality and Synthesis* (Durham: Duke University Press, 2002).

7. Mao Zedong 毛泽东, *Jingji wenti yu caizheng wenti* 经济问题与财政问题 (On Economic and Financial Questions) (Dongbei shuju, 1948), 41.

8. See here for instance the preface in Zhongguo xumu shouyi xuehui zhu chuanranbing zhuanye yanjiuzu 中国畜牧兽医学会猪传染病专业研究组, *Zhu de zhuyao chuanranbing jiqi fangzhi fangfa* 猪的主要传染病及其防治方法 (Main infectious diseases of pigs and their prevention methods) (Beijing: Caizheng chubanshe, 1956).

9. "Nongda shouyi shixi gongzuozhan—Zhaokai huiyi zongjie gongzuo jingyan 农大兽医实习工作站 召开会议总结工作经验 (Work station for intern veterinarians at the Agriculture University convened to summarize work experiences)," *Renmin Ribao*, July 8, 1950.

10. Jiangxi sheng xumu shouyi gongzuozhe xiehui 江西省畜牧兽医工作者协会 and Jiangxi sheng nongyeting zhongshouyi shiyansuo 江西省农业厅中兽医实验所, *Jiangxi minjian shouyi zhenliao ji chufang huibian* 江西民间兽医诊疗及处方汇编 (Compilation of folk veterinary diagnosis and formulae in Jiangxi) (Nanchang: Jiangxi renmin chubanshe, 1956), preface.

11. Zhou Enlai 周恩来, "Guowuyuan guanyu jiaqiang minjian shouyi gongzuo de zhishi 国务院关于加强民间兽医工作的指示 (1956)," in *Ningxia shouyizhi* 宁夏兽医志 (The Gazetteer on Veterinary Medicine in Ningxia), ed. Zhou Shengjun 周生俊 (Ningxia: Ningxia renminchubanshe, 2012), 341–43.

12. The Conference of Folk Veterinary Medicine (*Minjian shouyi huiyi* 民间兽医会议) that took place in September 1956 and discussed the guidelines for veterinary medicine valued the veterinary knowledge among local practitioners in a similar fashion. Participants of that conference were well-known veterinary doctors such as Yu Chuan 于船, Lin Xianhai 林先海, Guo Yongjiang 郭永江, Chang Bingyi 常秉彝, Wang Daofu 王道福, and Wang Liang 王亮.

13. Huangfu Yin 皇甫垠, "Rexin wei qunzhong fuwu de Zhongshouyi—Wang Daofu 热心为群众服务的中兽医——王道福 (The Chinese veterinary doctor enthusiastically serving the masses—Wang Daofu)," *Renmin Ribao*, January 27, 1957.

14. The journal was founded in 1957 by Yang Hongdao 杨宏道. In 1963, it was renamed *Materials on the Technology of Chinese Veterinary Medicine* (*Zhongshouyi keji ziliao* 中兽医科技资料), and in 1980 *Journal of Chinese Veterinary Medicine* (*Zhong shouyixue zazhi* 中兽医学杂志).

15. Xiong Dashi 熊大仕, "Guanyu xumu shouyixue de liangge wenti—Xiong Dashi de fayan 关于畜牧兽医学的两个问题——熊大仕的发言 (Two questions on animal husbandry veterinary medicine—Xiong Dashi's statement)," *Renmin Ribao*, March 25, 1957.

16. Wang, a specialist in economic and agricultural history, had studied at the Technical University of Munich and Paris during the 1920s and 1930s. After returning to China in 1946 he became professor of agricultural economy at Beijing University, and in 1949 professor of economy at Beijing Agricultural University. From 1952 to 1980 he was head of the library of the latter university, where he specialized in research on the history of agriculture.

17. Such an assessment is however by no means uniform in the large bulk of CVM literature in that era. A rare exception is a 1959 publication on veterinary acupuncture and moxibustion calling these methods scientific technologies (kexue jishu 科学技术) that result from a millennia-long struggle against veterinary diseases. See Zhongguo nongye kexueyuan zhongshouyi yanjiusuo 中国农业科学院中兽医研究所, *Zhongshouyi zhenjiuxue* (Acupuncture and moxibustion in Chinese veterinary medicine) 中兽医针灸学 (Beijing: Nongye chubanshe, 1959), 3.

18. Wang Yuhu 王毓瑚, *Zhongguo xumushi ziliao* 中国畜牧史资料 (Collection of Materials on the History of Animal Husbandry in China) (Beijing: Kexue chubanshe, 1958), 1–2.

19. Jiangxi sheng nongye kexue yanjiusuo zhongshouyi yanjiushi 江西省农业科学研究所中兽医研究室, *Zhu chuanqibing zhenjiu liaofa* 猪喘气病针灸疗法

(Acupuncture and Moxibustion to Treat Swine Asthma) (Nanjing: Jiangsu renmin chubanshe, 1959).

20. Jiangxi sheng nongyeting zhongshouyi shiyansuo 江西省农业厅中兽医实验所, *Shiyong shouyi zhenjiuxue* 实用兽医针灸学 (Nanchang: Jiangxi renmin chubanshe, 1958). This booklet was also thought to be presented at the Sino-Soviet Conference on Technological Cooperation in Veterinary Medicine (*Zhong-Su liangguo xumu shouyi jishu hezuo huiyi* 中苏两国畜牧兽医技术合作会议), see preface.

21. "Quanguo Zhong-Xi shouyi yanjiu gongzuo huiyi queding—jiaqiang Zhong-Xi shouyi de tuanjie hezuo—jiasu yanjiu zuguo shouyixue yichan 全国中西兽医研究工作会议确定 加强中西兽医的团结合作 加速研究祖国兽医学遗产 (National conference on research in Chinese and Western veterinary medicine decides to strengthen the united cooperation of Chinese and Western veterinary medicine in order to accelerate the research of the motherland's heritage in veterinary medicine)," *Renmin Ribao*, February 11, 1959.

22. See here Nongyebu guanyu Zhongshouyi "caifeng" he bianji Zhongshouyi yaowuzhi de tongzhi 农业部关于中兽医"采风"和编辑中兽医药物志的通知 (A notification of the Ministry of Agriculture regarding the Chinese veterinary medicine on "collecting local material" and compiling Chinese veterinary medicine pharmaceutical treatises)," a directive published by the Ministry of Agriculture, February 5, 1959. Here taken from Zhou Shengjun, *Ningxia shouyi zhi*, 13.

23. *Guowuyuan guanyu minjian shouyi gongzuo de jueding* 国务院关于民间兽医工作的决定 (The State Council's Decision on Folk Veterinary Work), in "Hunan sheng renmin weiyuanhui zhuanfa guowuyuan guanyu minjian shouyi gongzuo de jueding 湖南省人民委员会转发国务院关于民间兽医工作的决定," *Hunan zhengbao*, no. 9 (1963): 4–6.

24. Yu Benyuan 喻本元 and Yu Benheng 喻本亨, *Chongbian jiaozheng Yuan Heng liaoma niu tuo jing quanji* 重编校正元亨疗马牛驼经全集 (Beijing: Nongye chubanshe, 1963), "Chuban shuoming 出版说明 (Explanations on the publication)." The fight against superstition is not mentioned in the 1957 version of the *Therapeutic Treatise*.

25. Yu Benyuan and Yu Benheng, *Chongbian jiaozheng Yuan Heng liaoma niu tuo jing quanji*, 534–35.

26. See here Wan Guoding 万国鼎, "Gu nongshu de zhengli he chuban 古农书的整理和出版 (Editing and publishing ancient agricultural books)," *Renmin Ribao*, April 2, 1957. See also the review of the Zhujing daquan that argues quite similarly: Liu Xinhuai 刘新淮, "Zhujing daquan de pingjia jiqi fajue zhengli jingguo《猪经大全》的评价及其发掘整理经过 (Evaluation of the "Complete Pig Classics" and its process of excavation and editing)," *Zhongshou yiyao zazhi*, no. 6 (1985): 59–60. Francesca Bray has also shown in this context that modern China was able to build on a wide range of agricultural treatises (*nongshu*) that were not necessarily of elitist character, but also intended to transmit technological knowledge to farmers, such as those written and compiled in the context of encouraging agriculture (*quannong*), see Francesca Bray, "Chinese Literati and the Transmission of Technological Knowledge: The Case of Agriculture," in *Cultures of Knowledge: Technology in Chinese History*, ed. Dagmar Schäfer (Leiden: Brill, 2012), 299–326.

27. See here the following exemplary sources published during the 1950s: Lu Siman 陆思曼, *Jiachu siliao peizhi jiben zhishi jianghua* 家畜饲料配制基本知识讲话 (Addressing basic knowledge of food preparation for livestock) (Shanghai: Kexue jishu chubanshe, 1957); *Sulian jiachu fanyu de xianjin lilun he jingyan* 苏联家畜繁育的先进理论和经验 (Advanced Theory and Experiences of the Soviet Union's Livestock Breeding) (Beijing: Caizheng jingji chubanshe, 1956); Liu Ruisan 刘瑞三 and Shen Yancheng 沈延成, *Jiachu siyang* 家畜饲养 (Breeding livestock) (Shanghai: Shanghai kexue puji, 1958); Xie Guoxian 谢国贤 and Tao Lüxiang 陶履祥, *Jiachu binglixue zonglun* 家畜病理学总论 (General Discussion on Livestock Pathology) (Shanghai: Shangwu yinshuguan, 1951); Xie Yutong 谢裕通, *Jiachu yibing shouce* 家畜疫病手册 (Handbook on livestock plagues) (Shanghai: Zhonghua shuju, 1951); Fujian sheng nongye ting ed. 福建省农业厅, *Yang zhu siliao duo de hen* 养猪饲料多得很 (Raising pigs with fodder on a large scale) (Fuzhou: Fujian renmin chubanshe, 1959); Fujian sheng nongye ting ed. 福建省农业厅, *Yangzhu shouce* 养猪手册 (Handbook on keeping pigs) (Fuzhou: Fujian renmin chubanshe, 1957); *Zhongguo xumu shouyi xuehui* ed. 中国畜牧兽医学会, *Zuguo youliang jiachu pinzhong* 祖国优良家畜品种 (The motherland's fine livestock breeds) (Beijing: Kexue chubanshe, 1956).

28. Similar cases can be found in: Jiangxi sheng xumu shouyi gongzuozhe xiehui and Jiangxi sheng nongyeting zhongshouyi shiyansuo, *Jiangxi minjian shouyi zhenliao ji chufang huibian*.

29. See for example Heilongjiang sheng nongyeting xumu shouyichu ed. 黑龙江省农业厅畜牧兽医处, *Minjian shouyi yanfang* 民间兽医验方 (Harbin: Heilongjiang renmin chubanshe, 1958); Zhongguo renmin jiefangjun shouyi daxue 中国人民解放军兽医大学, *Zhongshouyi yanfang ji* 中兽医验方集 (Collected Recipes in Chinese Veterinary Medicine) (Beijing: Jiefangjun shouyi daxue, 1972); or the collection *Quanguo zhongshouyi jingyan xuanbian* 全国中兽医经验选编 (Selected experiences of Chinese veterinary medicine) (Beijing: Kexue chubanshe, 1977).

30. See Anhui sheng nongyeting xumu shouyichu 安徽省农业厅畜牧兽医处, *Anhui zhongshouyi yaofang huibian* 安徽中兽医药方汇编 (Collected recipes of Chinese veterinary medicine in Anhui) (Hefei: Anhui renmin chubanshe, 1958), preface.

31. The insistence on *jingyan* as a source of knowledge in the collection of recipes had already been emphasized in Yu Yan's critique of theoretical Chinese medicine in the 1920s. See Lei, "How Did the Chinese Medicine Become Experiential? The Political Epistemology of Jingyan."

32. Chen Huan 陈焕, "Yong niushi wei zhu de banfa hao 用牛屎喂猪的办法好 (The method of using cow dung to feed pigs is good)," in *Fazhan shengxu shengchan de jingyan* 发展生畜生产的经验 (*Nongye shengchan jingyan congshu* 农业生产经验丛书), ed. Zhonggong Guangdong sheng weiyuanhui bangongting 中共广东省委员会办公厅 (Guangzhou: Guangdong renmin chubanshe, 1958), 36–40. The argument was that, compared to pigs, cows have a less efficient digestive system, as they don't absorb all the nutriments in their fodder.

33. The justification to do so was provided in the 1956 directive of the State Council on veterinary medicine. See Zhou Enlai, *Guowuyuan guanyu jiaqiang minjianshouyi gongzuo de zhishi*.

34. Zhubing fangzhi bianxiezu, ed. 猪病防治编写组, *Zhubing fangzhi* 猪病防治 (Prevention and treatment of swine diseases) (Shanghai: Shanghai renmin chubanshe, 1972), 147–49.

35. Jiangxi sheng nongyeting 江西农业厅 and Shouyi fangzhizhan 兽医防治站, *Tubing fangzhifa* 兔病防治法 (Prevention and treatment of rabbit diseases) (Beijing: Nongye chubanshe, 1959), 61.

36. The functionality of nerves in this context is explained by a reference to the experiments of the Russian physiologist Ivan Pavlov (1849–1936). See Jiangxi sheng nongye kexue yanjiusuo zhongshouyi yanjiushi, *Zhu chuanqibing zhenjiu liaofa*, 28.

37. The formula stems from Luquan County in Yunnan Province 云南禄劝县. See Nongyebu xumu shouyiju 农业部畜牧兽医局, *Zhongshouyi yanfang huibian (1956 nian quanguo minjian shouyi zuotanhui ziliao)* 中兽医验方汇编 (1956 年全国民间兽医座谈会资料) (A Collection of tested recipes of Chinese veterinary medicine: materials from the 1956 National Forum of Folk Veterinary Medicine) (Beijing: Nongye chubanshe, 1957), 143–46.

38. As noted in the *Yuan Heng Therapeutic Treatise of Horses*, see Yu Benyuan and Yu Benheng, *Yuan Heng liaoma ji*, 1.

39. Cf. the first chapter in Zhubing fangzhi bianxie zubian, *Zhubing fangzhi*.

40. Qinbing fangzhi bianxiezu 禽病防治编写组, *Qinbing fangzhi* 禽病防治 (Prevention and treatment of poultry diseases) (Shanghai: Renmin chubanshe, 1975). See also the preface in the 1972 edition of Zhongguo nongye kexueyuan zhongshouyi yanjiusuo 中国农业科学院中兽医研究所, *Zhongshouyi zhenduanxue* 中兽医诊断学 (Diagnostics in Chinese Veterinary Medicine) (Beijing: Nongye chubanshe, 1972) (originally published in 1962).

41. For the role of barefoot doctors in Mao-era China see the comprehensive study by Fang Xiaoping, *Barefoot Doctors and Western Medicine in China*.

42. "Hanjiang nongye xuexiao duo kui hao sheng peiyang nongye jishu renyuan. Jiangsu sheng Yangzhou zhuanqu geming weiyuanhui huo de diaocha baogao 邗江农业学校多快好省培养农业技术人员。江苏省扬州专区革命委员会的调查报告 (Hanjiang Agricultural School trains agricultural technical staff more, faster, better and more economical. Investigation report of Special District Revolutionary Committee in Yangzhou City in Jiangsu Province)," *Renmin Ribao*, October 26, 1968.

43. "Shixing shengzhu hezuo baoxian zhidu cujin yang zhu shiye da fazhan. Guangdong sheng Dongguan xian Dalang gongshe Huangcaolang dadui zai Mao zhuxi guanyu fazhan yang zhu shiye zhongyao zhishi zhiyin xia, da yang qi zhu, cujin le nongye shengchan da fazhan 实行生猪合作保险制度促进养猪事业大发展。广东省东莞县大朗公社黄草朗大队在毛主席关于发展养猪事业重要指示指引下，大养其猪，促进了农业生产大发展 (Implementing the Cooperative Insurance System for raising pigs to promote the great development of the pig industry. Under Chairman Mao's important guidance on developing the pig-keeping industry, the Huangcaolang Production Brigade of the Dalang Commune in Dongguan county, Guangdong Province, eagerly raised their pigs, boosted the great development of agricultural production)," *Renmin Ribao*, May 16, 1969.

44. "Jiji peiyang chijiao shouyi duiwu 积极培养赤脚兽医队伍 (Vigorously train barefoot veterinarians)," *Renmin Ribao*, December 12, 1975.

45. "Reqing peiyang chijiao shouyi 热情培养赤脚兽医 (Passionately training barefoot veterinarians)," *Renmin Ribao*, February 1, 1976. An overview on reports on barefoot veterinarians in the *People's Daily* in the mid-1970s tells how these efforts apparently contributed to reducing mortality of pigs, treating several hundreds of animals within a short time span, or raising productivity. See, for example, "Jiji fangzhi zhuyibing 积极防治猪疫病 (Vigorously preventing pig plague)," *Renmin Ribao*, July 2, 1975.

46. Du Nianxing 杜念兴, *Shouyixue dayi* 兽医学大意 (An Overview of Veterinary Medicine) (Nanjing: Xumu shouyi tushu chubanshe, 1957).

47. As reported in Yunnan sheng nongyeju 云南省农业局 and Xumu shouyi kexue yanjiusuo 畜牧兽医科学研究所, *Zhongshouyi jingyan xuanji (shang)* 中兽医经验选辑 (上). (Selected experiences of Chinese veterinary medicine) (no publisher indicated, 1973).

48. Jiangxi sheng nongye kexue yanjiusuo zhongshouyi yanjiushi, *Zhu chuanqibing zhenjiu liaofa*.

49. See for example Hebei Zhongshouyi xuexiao 河北中兽医学校, Beijing budui houqinbu weishengbu 北京部队后勤部卫生部, Dongbei nongken daxue 东北农垦大学, Zhou xian shouyiyuan 涿县兽医院, *Zhongshouyi shouce* 中兽医手册 (Manual of Chinese Veterinary Medicine) (Beijing: Nongye chubanshe, 1975), 315.

50. Sanyuan xian kexue jishu weiyuanhui 三原县科学技术委员会 and Sanyuan xian kexue jishu xiehui 三原县科学技术协会, *Zhuwen de zazhen zhiliao* 猪瘟的扎针治疗 (Acupuncture treatment of swine diseases) (Xi'an: Shaanxi renmin chubanshe, 1960), 11.

51. Jiangsu sheng nongye kexue yanjiusuo geming weiyuanhui 江苏省农业科学研究所革命委员会, *Zhongshouyi zhenliao xuanbian* 中兽医诊疗选编 (Selected cases of Chinese veterinary medicine diagnosis and treatment) (Nanjing: Jiangsu sheng geming weiyuanhui, 1970), annex.

52. Cf. here Li Qingbin 李庆斌, "Wei geming dang hao chijiao shouyi 为革命当好赤脚兽医 (Barefoot veterinarians for the revolution)," *Xiandai xumu shouyi*, no. 4 (1975): 10–16; as well as Zhang Shengwen 张声文, "Chijiao shouyi wei renmin 赤脚兽医为人民 (Barefoot veterinarians for the people)," *Guangxi nongye kexue*, no. 5 (1976): 40–41; Zhubing fangzhi bianxiezu ed., *Zhubing fangzhi;* Revolutionary Health Comittee of Hunan Province, ed., *A Barefoot Doctor's Manual* (London: Routledge, 1977).

53. "Jiji fangzhi zhu yibing," *Renmin Ribao*, July 2, 1975.

54. Judith Farquhar, *Knowing Practice: The Clinical Encounter of Chinese Medicine* (San Francisco: Westview Press, 1994), 147–74.

55. Cf. Zhongguo kexueyuan dongwu yanjiusuo 中国科学院动物研究所, *Zhubing fangzhi shouce* 猪病防治手册 (Pamphlet of preventing swine diseases) (Beijing: Kexue chubanshe, 1979), 25–30. This is similar to the principle of *bianzheng lunzhi* 辨证论治 (pattern differentiation and treatment determination). According to Volker Scheid it characterizes the essence of clinical practice in Chinese medicine, but only appeared during the 1950s. See Scheid, *Chinese Medicine in Contemporary China*, 200–37, as well as Volker Scheid, "Convergent Lines of Descent: Symptoms,

Patterns, Constellations, and the Emergent Interface of Systems Biology and Chinese Medicine," *East Asian Science, Technology and Society* 8, no. 1 (2014): 107–39.

56. Ralph Croizier, *Traditional Medicine in Modern China: Science, Nationalism, and the Tensions of Cultural Change* (Cambridge: Harvard University Press, 1968); Lei, *Neither Donkey nor Horse*. See also the entry on Academic Societies of Chinese Veterinary Medicine (*Zhongshouyi xueshu tuanti* 中兽医学术团体) by Xie Zhongquan 谢仲权 and the entry on Chinese Veterinary Research Institutions (*Zhongshouyi yanjiu jigou* 中兽医研究机构) by Meng Xiansong 孟宪松, in *Zhongguo nongye baike quanshu: Zhongshouyi juan* 中国农业百科全书：中兽医卷 (Chinese agriculture encyclopedia: volume on Chinese veterinary medicine) (Beijing: Nongye chubanshe, 1991), 384–85.

Chapter Six

Re-shuffling Science in the Reform Era

The preceding three case studies have shown that raising productivity was a crucial concern in the era of socialist construction. Facing a constant shortage of capital but a surplus of labor—a problem that had persisted since the imperial era, as Mark Elvin shows in his discussion on the high equilibrium trap[1]—China embarked on a different path of modernization. When reading Mao Zedong's seminal texts on political economy—among them *Economic and Financial Problems in the Anti-Japanese War* (*Kang-Ri shiqi de jingji wenti he caizheng wenti* 抗日时期的经济问题和财政问题, 1942), *The Ten Great Relationships* (1956), and *The Critique of Soviet Economics* (*Du shehuizhuyi zhengzhi jingjixue pizhu he tanhua* 读社会主义政治经济学批注和谈话, 1960)—one discovers a number of approaches to cope with the labor surplus and lack of arable land. According to Mao, labor can compensate for any shortage of land, capital, and technology, an idea that had already been discussed in the Yan'an years.[2] The drain of resources caused during the Great Leap Forward, when agricultural production was forced to finance industrialization, was supposed to be overcome with local self-reliance and small-scale mutual aid, combined with local innovations. In the case of agriculture, as shown earlier, this meant to prefer half-mechanization over the indiscriminate introduction of heavy machinery. In steel production, individual innovations created on the shopfloor by the industrial worker were valorized as knowledge gained from practice to improve collective productivity, even if capital-intensive machinery was lacking. With regard to veterinary medicine, the emphasis on indigenous knowledge and local drugs strengthened a combination of Chinese and Western medicine that was not a unified system, but a locally contingent one.

Therefore, the preceding chapters have tried to locate various forms of knowledge in the Mao era, despite the fact that sometimes they may appear weird or unorthodox. Once believed to push forward innovations in agricultural and industrial production, Great Leap Forward practices of using nodular cast iron, veterinary acupuncture, or ball-bearings produced from ceramics and acorns are often overlooked or despised as anti-scientific practices today, especially in publications that praise the return of science after the end of the Cultural Revolution.[3] The negative assessment of the discourse of science in Mao-era China was, to a large extent, the result of the radical transformation of science and technology at the beginning of the reform era. Accompanied by numerous reform measures right after the official end of the Cultural Revolution in 1976, this transformation called for the re-professionalization of science and technology by putting an end to political influence and mass voluntarism. Political discourse in post-revolutionary China has emphasized the impact of this transformation on the successful economic growth in the recent decades, while dismissing Mao's notion of mass science and technology as insignificant or useless. Yet, as we are going to argue in the following, despite the re-professionalization that was oriented to the science of the so-called modern knowledge, knowledge production of the Mao era has had a far-reaching impact on China's science until today.

EPISTEMIC RUPTURE OR CONTINUITY? SCIENCE AND TECHNOLOGY AS PRIMARY PRODUCTIVE FORCES

In March 1978 the new leader Deng Xiaoping opened the National Science Conference (*Quanguo kexue dahui* 全国科学大会) with an address that intended to restore the authority of professional science and scientists. Deng openly criticized the Gang of Four and Lin Biao for interrupting scientific development and persecuting intellectuals and scientists, depicting their radical policies as an impediment to China's economic progress.[4] At the end of the Cultural Revolution the Gang and their accomplice Lin were seen as the major culprits of the anti-scientific and anti-intellectual situation that jeopardized the nation's future.[5] According to Deng, the lesson of the Cultural Revolution was that the state should improve the people's living standards by achieving economic growth. Science and technology were now considered the primary productive forces (*kexue jishu shi diyi shengchanli* 科学技术是第一生产力), and contrary to the claims of the Gang of Four, the mental worker (or intellectual)[6] had already become a member of the working class. The future of socialism was now to develop strengths in each field by focusing on expert knowledge, he pointed out optimistically.[7] In post-1978 historiography of the People's Republic his speech is often seen as a breakthrough in science policy. In contrast to the situation during the Great Leap Forward and the Cultural Revolution, politics was no longer in command, and the autonomy of scientists was restored as a central part of the official policy of the Four Modernizations (*sige xiandaihua* 四个现代化).[8]

Recent publications on the history of science have called for more nuanced approaches to (the discourse of) science in Mao-era China, which try to tease out its epistemic multiplicity and scientific progress in various fields, ranging from military science to agriculture and industry. Joshua Eisenman and Sigrid Schmalzer argue that the contributions of nation-wide agricultural research and the extension system have been largely underestimated.[9] Commune-based small teams worked with agricultural technicians to develop, test, and disseminate new technologies, characterized by vertical communication patterns that allowed speedy dissemination, while the semiautonomous character of the teams ensured successful testing and thus knowledge production. An important element of this science policy was the continuous emphasis on experiment as a valuable method to create scientific knowledge. In the mid-1970s, before the death of Mao, the innovative character of the agricultural scientific experiment movement was praised by the U.S. National Academy of Sciences Plant Science delegation. After a visit to China it emphasized the significance of taking the "accumulated knowledge of generations of peasants' experience" seriously.[10] Likewise, the editors of *China: Science Walks on Two Legs* (1974) recognized the success of de-professionalized science and viewed the mass-line as a valid epistemological principle, a view that was possibly shaped by their leftist critical attitudes toward scientific practices in the United States and by the reports in China's official media such as the *Peking Review* and the *China Pictorial*. These official publications that were available in English translation privileged applied knowledge over theoretical science as the most significant source of innovation.[11]

The need to tease out the positive aspects of knowledge production in the Mao era is reiterated in *Mr. Science and Chairman Mao's Cultural Revolution: Science and Technology in Modern China* edited by Chunjuan Nancy Wei and Darryl Brock. They warn against blindly accepting the Cold War dichotomy of a liberal West protecting the autonomy of science vs. an interventionist socialism persecuting bourgeois experts. This dichotomy has not only privileged a Western notion of science as universal, but also shrouded the heterogeneous character of scientific knowledges in Chinese society. This edited volume argues that even during the Cultural Revolution a number of important innovations can be identified. For example, the first hydrogen bomb test was completed in 1967, integrated circuits were produced in 1968, and the first satellite was launched in 1970. The 1960s saw the construction of a large radio telescope in time to observe the 1968 period of intense solar activity, an expedition to Mount Everest was accomplished, insulin synthesized, and the world's first synthetic benzene plant built.[12] Such breakthroughs show that at least after the first years of intense political fights in the Cultural Revolution there was visible progress in science and technology, urging the historian to acknowledge and critically analyze the much more complex picture of the Cultural Revolution.

The motivation of this book was not to paint a similarly optimistic picture of the Mao era. Rather, we want to argue that the assumption of an epistemic rupture in 1978, which has been lauded for restoring the autonomy of science and elevating it to the status of the primary productive forces, tends to conceal the continuities of the ideas of science and technology in the Mao era into today's China and thus fail to explain their lasting impact. On the one hand, the decision by Deng started a profound transformation of Chinese society. It has strengthened today's legitimacy of the CCP by declaring his progressive achievements as a liberation from the ideology of permanent revolution (*buduan geming* 不断革命). On the other hand, the assumption that the authority of innovating new knowledge was restored to the expert after 1978 implies that farmers and workers are moved into the second rank, that is, becoming mere passive receptors of knowledge. As detailed by Joel Andreas, the return of the knowledge production as a hierarchical and exclusive social network contributed to the emergence of a red elite that has reproduced the convergence of political and educated elites.[13]

However, if knowledge is produced hierarchically and disseminated in a top-down fashion, today's belief in the omnipotence of science and technology across all social classes would primarily depend on the effectiveness of propaganda. Science and technology, however, are more than mere propaganda. The question is whether Deng's 1978 speech can be seen as a political maneuver that simply bid farewell to the rhetoric of class struggle, or whether it could amount to an epistemic rupture in knowledge production. We may start by assessing how deep this rupture actually is.

In February 1978—after Deng Xiaoping had been put in charge of education, science, and technology—a new Chinese Academy of Social Science (CASS) was established under the leadership of Deng's associate Hu Qiaomu 胡乔木 (1912–92).[14] Its vice-president-to-be Yu Guangyuan 于光远 (born in 1915) defined this move as essential to the renaissance of social science research, claiming that the situation had improved compared with that in the Cultural Revolution: scientists and scholars now enjoyed not only social prestige but also freedom and autonomy in academic discussion.[15] Deng declared in his speech at the opening ceremony of the National Science Conference that intellectuals were part of the socialist working class who contributed to the productive forces. He pointed out that "a mammoth force of scientific and technical personnel who are both red and expert" had to be assembled to overcome China's backwardness.[16]

Sleeboom-Faulkner points out in her history of the CASS that its leaders of the first generation pursued the aim of promoting the status of the intellectuals and acknowledging their important role in the reform.[17] To incorporate the intellectuals into the working class resonated with Zhou Enlai's demands of 1956 (the moderate position that pursued technocratic visions in the Mao era) as well as with Lu Dingyi's ideal of the mass-line (the position of the radical left). The swinging between both positions might be attributed to a general sense of caution in political issues shortly after the Cultural Revolution (or as a conciliatory move by Deng toward the leftists), but its significance lies in its contribution to the revitalization of science and scientists.[18] It negated the class struggle character of science and technology and saw that both were universal in nature and have the primary purpose of creating wealth. As a consequence, mental workers were again (as in January 1956) re-categorized into workers in the Marxist sense (*laodongzhe* 劳动者). By doing so, party secretaries in research institutes were forced to grant greater autonomy to the researchers, and political education in state research institutions was reduced. At the same time, the differentiation between bourgeois and proletarian science disappeared, and the role of the masses in producing knowledge for modernization diminished. For instance, the mathematician Wang Zikun 王梓坤 (born in 1929), in a 1979 article, mentioned the Chinese people (*renmin*) only briefly as a hard-working, intelligent, and brave people who made great achievements in the ancient past (the past thus not being anti- or pre-scientific, i.e., mired in superstition); now the socialist state produces outstanding scientific theoreticians and engineers that are necessary for modernization. The people are mentioned in this text merely in passing and this seems to be either done for mere rhetorical reasons, or as reminiscence to the past ideal of the mass-line. For Wang, to love the people, to love truth, and to love labor (*reai renmin, reai zhenli, reai laodong* 热爱人民，热爱真理，热爱劳动) are essential preconditions of science work, and he reminds that this idea is not alien to the elite scientist.[19] In other words, being an expert no longer constitutes a problem because class struggle has ceased to exist and the experts are educated in the socialist spirit.

Wang's reassurances notwithstanding many scientists felt a lack of theoretical support for scientific research, which was the major reason for resuming the publication of the *Newsletter of Natural Dialectics* (*Ziran bianzhengfa tongxun* 自然辩证法通讯) that had stopped earlier in 1966. After the purge of the Gang of Four, a conference in March 1977 discussed the need for a journal that helps scientists and technicians to grasp natural dialectics. Three of its participants—Yu Guangyuan, Li Chang 李昌 (1914–2010),[20] and Qian

Sanqiang 钱三强 (1913–92)[21]—requested Fang Yi 方毅 (1916–97), the vice premier and president of the Chinese Academy of Sciences (CAS) from 1979 to 1981,[22] and Deng Xiaoping in a letter to grant permission in October of that year. The first issues published in 1978 concentrated on justifying the republication of the journal by pointing to the wrong-doings of the Gang of Four that was condemned for having used natural dialectics to intervene with the work of natural scientists.[23]

The protection from political intervention and the return to academic freedom were justified in the January 1979 issue of the *Newsletter* with a reference to 1950's Lysenkoism.[24] The rebuttal of Lysenkoism prominently discussed in this issue was more than a simple refutation of a wrong scientific idea or attitude. It was a reminder that the success of a scientific theory can be assured only if one maintains Mao's insight that epistemology has to be grounded in the social realm, i.e., combining scientific experiments with dialectical thinking in order to keep the unity of mental and manual labor, as Wang Zikun states in a contribution to the *Red Flag*. Neither can be separated from the other, and even though scientific progress is becoming more and more difficult, it is at the same time true that science is no longer a matter of single, outstanding individuals such as Newton or Edison. It requires the state or even multinational collectives who are able to plan, organize, and finance research.[25]

The *Newsletter of Natural Dialectics* was republished precisely at that time when the call for greater academic freedom was getting into conflict with the Maoist epistemology of science. Its task was to come up with a science theory whose function was to identify the permissible limits of research. In the course of the Campaign against Spiritual Pollution (*qingchu jingshen wuran*清除精神污染), the *People's Daily* identified in December 1983 areas of science immune to pollution, among which were science and technology, scientific debate, and the freedom to choose research projects.[26] There were virtually no forbidden zones in academic research, as the historian and Marxist theoretician Hu Sheng 胡绳 (1918–2000) wrote in a contribution to the *Red Flag* in 1986 on the occasion of the thirtieth anniversary of the Hundred Flowers campaign. In the past Marxism had not been understood properly, which made life difficult for the intellectual worker, he argued. Since class struggle had lost its significance, it was no longer necessary to impose Marxist ideology as this would be detrimental to the newly cherished ideal of academic freedom.[27] Hu Sheng thus reiterated the idea that political intervention in scientific work was an obstacle in the development of science, a point raised by the president of the Chinese Academy of Sciences, Fang Yi, in 1980.[28]

In 1982 the *Red Flag* brought out an article summarizing the experiences of the ongoing reform and urged upholding the long-cherished ideal of the past, that is taking the mass-line in pursuing reform was imperative because the reform cannot be administered from above.[29] It was then the task of scientists to elicit this support by explaining to manual workers what purpose their research had, and how this could improve the living standard of the population and increase economic growth.[30] In other words science should not be an undertaking of only the elite, but one that concerns the whole of society: the top-down structure of science popularization was reestablished, although science was no longer a monolithic existence any more. The remaining part of this chapter is going to show how elitist scientists—despite enjoying a new social and political status—still had to engage with knowledge that does not conform with the universal and unified notion of science with its presumed objectivity and rationality, and in some cases, such conflict could take place among the science experts themselves.

THE FUTURE OF LOCAL KNOWLEDGE TODAY

As discussed earlier, the discourse of science and technology neither lost its ideological package during the first years after the Cultural Revolution, nor was the authority of the scientific expert restored immediately. The political interventions both during and after the Mao era notwithstanding, science and technology helped to establish a technocratic system in which science possesses an image of security and reliability in solving problems as well as contributing to economic development, as many monographs on science and science policy in the PRC attest.[31] Until today, this belief is largely unshaken, contrary to the situation in a number of Western countries where the disasters at Chernobyl and Fukushima shook the belief in the omnipotence of science severely.[32] Yet, it would be too simple to explain the persistence of such belief by a lack of similar catastrophes in China, or by the pessimist view of Anthony Giddens who saw trust in science as caused by the circumstance that modern man needs to have faith in expert systems whose details are unintelligible to him.[33]

On the contrary, the belief in science and its system in the Chinese context can be traced to two sources. One is the authority of science established in the Republican era as a result of the May Fourth Movement and the ensuing state-building of the Nationalist government. The other is the conscious and particularly extensive dissemination of science in the Mao era. The latter case can at first sight be explained, to some extent, by the power of propaganda. To account for its persistence in the post-Mao age, we must also acknowledge the significance of knowledge production and dissemination as social practice. Knowledge production is an all-encompassing social process, so defines Mao Zedong's in his text *On Practice*. It is characterized by a large degree of pragmatism that takes local resources and local initiatives seriously, thereby strengthening the view that the dissemination of science and technology does not happen in an exclusive top-down fashion. It is linked to those in the field and on the shopfloor who contributed their personal experiences and breakthroughs in experiments in order to achieve the Marxist ideal of a classless society.

As shown by Joel Andreas in his study on the "red engineers," social reality looked much more complicated. During the Cultural Revolution, the Marxist ideal was combined with the technocratic ideal of Saint-Simon to create the caste of the red engineers, who enjoyed both professional and political prestige. This might very well explain their social high standing especially in post-revolutionary China. Yet, we argue that the widespread belief in science and technology can only be understood when we view the call for removing the boundaries between mental and manual labor—an idea still shared by Deng Xiaoping in the 1990s[34]—not as mere political rhetoric, but rather as a real, influential issue concerning legitimacy and authority of knowledge. In the Mao era, the efforts to take into account bottom-up knowledge production rooted in pre-modern epistemics or popular religions allowed the emergence of obscure practices, such as white steel, animal acupuncture, *qigong*, or chicken-blood therapy, that were at some time viewed as pseudoscientific or even superstitious.[35]

Today they are more or less counted as part of China's cultural tradition, especially since the culture fever movement (*wenhuare* 文化热) of the 1980s. Largely understood in an essentialist fashion, culture was supposed to nurture a sense of patriotic pride; and the recourse to tradition went hand in hand with the questioning of the universality of so-called Western science by emphasizing alterity.[36]

Their origin, however, goes back to the pre-1978 era when the Maoist epistemology of science was still prevailing. In the 1980s, expert-scientists continued to justify alternative knowledges by pointing to experience as a valid epistemology. When the father of China's rocket program Qian Xuesen 钱学森 (1911–2009)—an acclaimed scientist who had been educated and worked in the United States—propagated parapsychological ideas such as special (psychic) abilities (*renti teyi gongneng* 人体特异功能), he proposed the foundation of a new discipline called "somatic science" (*renti kexue* 人体科学) built on Chinese medicine and *qigong*.[37] As their origins were quite "naturally" to be found in the Chinese civilization, somatic researchers claimed that they had a general and decisive advantage over Western medicine because their *qigong* was a systematic method that had been tried and tested out for over more than 5,000 years. In other words, the thousands of years of experience turned it and other somatic practices into valuable and timeless practices. The social epistemology with its focus on experiential science (*jingyan kexue* 经验科学) thus still persisted.[38]

Critics of the naïve belief in Chinese medicine such as the quantum field theorist and ardent believer in Marxism He Zuoxiu 何祚庥 (born in 1927)[39] still takes a seemingly equivocal position by defending *qigong* for its contribution to personal health while at the same time demanding the abolition of Chinese medicine. In the former case, *qigong* is a legitimate body technique due to its generally acknowledged effects, and in the latter case he argues that it is a theoretically weak discipline that cannot qualify as science. On the surface, his critique resembles the one once uttered by Yu Yunxiu in the fight against national medicine in the 1930s, but in fact their arguments are different, especially concerning Chinese medicine's relation to Chinese culture: He Zuoxiu argues that understanding CM as part of China's rich cultural heritage does not help to prove its scientific character. Defending the assumption of a unified science that does not know national borders, he asks how one could construe a Western and a Chinese medicine. In contrast to Yu, however, He Zuoxiu locates the problem in the reference to culture: today's prevailing cultural essentialism has caused some people to believe in the superiority of Chinese medicine vis-à-vis Western biomedicine but prevented them from analyzing both rationally.[40]

In other words, culture alone is insufficient to explain why certain traditional knowledges are preferred over others. Today's global discourse in the history of science has reacted to the growing recognition of non-Western forms of science and technology by replacing the notion of science with that of knowledge.[41] The scholar Lan Liji pointed out in the late 1990s that medicine can no longer be explained with reference to science exclusively, but has to be seen as a cultural phenomenon in an anthropological sense.[42] More forcefully Nie Jingbao argued that Chinese medicine has to be rescued from the heritage of the May Fourth Movement that has negated its existence due to the exclusive preference of Western medical science. In fact, medical science, hygiene, and health can never be separated from social or historical circumstances, and Chinese medicine is no exception, thus it deserves to be preserved as a distinct system of knowledge that should not be subjected to the category of Western scientism, so is Nie's argument.[43]

The call for a plural understanding of science by explicitly historicizing knowledge is not specifically Chinese. When the renowned Isis Current Bibliography in the History of Science removed in 2002 the category "pseudoscience" from its bibliography, publications in the disciplines of astrology and alchemy are now listed in the same categories with astronomy and chemistry.[44] This signals that the history of *the* science, which is defined by its quest for universal knowledge and impartial truth and which had been spread globally by the time of First World War, has been increasingly put into question. The dissolution of Western science's claim of universality might be viewed as a postcolonial move that intends to do justice to local science traditions. Peter Dear has shown in his reflections on the roots of modern science that scientific ideas, practices, and institutions are the result of historical contingency. While their increasing historicization has caused the category "science" to become slippery, the history of science has been pluralizing itself.[45] In the Chinese case, this took place when Mao's discourse on experiment helped to rescue traditional practices from obscurity and oblivion, yet without resulting in a fearsome "relativistic nightmare," as has been stated by Marwa Elshakry. Contrary to her pessimistic view that the "postmodern turn never really caught on in the Eurocentric history of science"[46] and that challenges to the singularity of Western science are hardly seen,[47] the examples in this book have shown that science in the Mao era was characterized by an epistemological flexibility to accept a plurality of knowledges shaped by the political and social transformations in that era—before culture became a central argument.

The preceding chapters have argued that the recourse to practice not only valorized the worker's and farmer's experiences, but also contributed to the emergence and dissemination of different knowledge systems. With the socialist state penetrating every level of society, this plurality was able to spread extensively. In other words, it was the epistemology of practice that enabled the creation of coexisting (and sometimes contradictory) knowledges in society, and less the cultural turn, which only later reinforced the positive and authentic image of past knowledge practices. While Hu Shi, following the Western model, declared at this time the persistent and pervasive existence of Mister More-or-Less, a personification of a national character, an obstacle to progress, China's search for an alternative way to modernity in the Mao era gave rise to knowledge production that did not necessarily follow a unified, universal, and implicitly Eurocentric notion of science. The temporary shift of scientific authority to the masses accommodated the inclusion of local ideas on science and scientific practices. Therefore we propose that it would be more productive to conceptualize the history of science in China as a history of knowledge, because this move can help us to steer away from Eurocentric conceptions of science and scientific thinking and to incorporate ideas and practices from different geographical spaces and epistemological traditions within China.[48]

The effects of learning science from the masses can still be observed in the contemporary era where, despite the dominance of advanced technology and natural science in modernization, plural knowledge epistemologies that are derived from imagined continuities and invented traditions coexist. In such a society, the state needs to be creative to accommodate and reconcile contradicting epistemologies. As our case studies show, the social epistemology with its emphasis on experience and practice included ordinary people into the community of knowledge production, which may not only bring out their creativity to deal with problems in local contexts, but also helped them to develop a sense of belonging to the state. An alienated populace would eventually weaken the state's science dissemination agenda in a society, where the 2003 official call for cultivating the outlook of scientific development (*kexue fazhanguan* 科学发展观) defined again science and technology as the primary path to wealth and power.[49] In the end, with this examination of the validation of local and practical knowledge(s) in the social turn of 1950s and 1960s China, we historicize the notion of science and tease out the strands of continuity rather than rupture between the pre-modern and the modern, as well as the revolutionary and the reformist eras.

NOTES

1. Elvin, "The High-Level Equilibrium Trap"; Isett, *State, Peasant, and Merchant on the Manchurian Frontier*.
2. Mark Selden, "Mao Zedong and the Political Economy of Chinese Development," *China Report* 24, no. 2 (1988): 125–39.
3. This is a typical and often heard argument in the years after 1978. For instance, a 1982 contribution to the *Hongqi* declared that due to the leftist tendencies in the Cultural Revolution the education and training of employees had suffered enormously. Their situation improved only after the start of the Four Modernizations, see Yuan Baohua 袁宝华, "Jiaqiang zhigong jiaoyu shi shixian sige xiandaihua de zhongda zhanlüe renwu 加强职工教育是实现四个现代化的重大战略任务 (Strengthening the Workers' Education Is a Great Strategic Mission of Achieving the Four Modernizations)," *Hongqi*, no. 5 (1982): 17–20.
4. A similar assessment has been made by Needham, see here Joseph Needham, "Science Reborn in China: Rise and Fall of the Anti-intellectual Gang," *Nature* 274 (1978): 832–34. For Deng's Speech at the Opening Ceremony of the National Science Conference, March 18, 1978, see *Quanguo kexue dahui wenjian* 全国科学会议文件 (*National Science Conference Documents*) (Beijing: Renmin chubanshe, 1978), 14–30; for an English translation see Deng Xiaoping, "Speech at Opening Ceremony of National Science Conference," *Peking Review*, no. 12 (1978): 9–18.

Christine Luk comes to a different conclusion in her dissertation, noting that the Cultural Revolution had a very negative impact on the Chinese Academy of Sciences, yet this did not necessarily have consequences for Chinese science in general because individual scientists could continue their research under new institutional headings. Christine Yi Lai Luk, *A History of Biophysics in Contemporary China* (Cham: Springer, 2015).

5. See here "Dongyuan qilai, jiasu shixian kexue jishu xiandaihua 动员起来，加速实现科学技术现代化 (Mobilizing everyone to speed up Scientific and Technological Modernization)," *Hongqi*, no. 7 (1977): 4.

6. Deng uses both terms here, declaring that "the difference between intellectuals and manual workers lies only in their different roles in the social division of labour (他们[知识分子]与体力劳动者的区别，只是社会分工的不同)." See *Deng Xiaoping wenxuan* 邓小平文选 (Beijing: Renmin chubanshe, 1993), Vol. 2, 89.

7. See here *Quanguo kexue dahui wenjian*. The idea to do so was already present in August 1977 when Deng Xiaoping called for the restoration of the State Science and Technology Commission that had been abolished during the Cultural Revolution. See here the discussion in Lyman Miller, *Science and Dissent in Post-Mao China: The Politics of Knowledge* (Seattle: University of Washington Press, 1996).

8. For an overview see Yang Dali, "State and Technological Innovation in China: A Historical Overview, 1949–89," *Asian Perspective* 14, no. 1 (1990): 91–112.

9. Schmalzer, *Red Revolution, Green Revolution*; Eisenman, *Red China's Green Revolution*.

10. American Plant Studies Delegation, *Plant Studies in the People's Republic of China: A Trip Report of the American Plant Studies Delegation* (Washington, D.C.: National Academy of Sciences, 1975), 118. See also Schmalzer, *Red Revolution, Green Revolution*; and American Insect Control Delegation, *Insect Control in the People's Republic of China: A Trip Report of the American Insect Control Delegation, Submitted to the Committee on Scholarly Communication with the People's Republic of China* (Washington: National Academy of Sciences, 1977).

11. See Science for the People, *China: Science Walks on Two Legs—A Report from Science for the People* (New York: Avon Books, 1974).

12. Han Dongping, "Rural Agriculture: Scientific and Technological Development during the Cultural Revolution," in *Mr. Science and Chairman Mao's Cultural Revolution*, 281–304. Darryl E. Brock provides an overview on how these achievements were perceived outside China, "The People's Landscape: Mr. Science and the Mass Line," in *Mr. Science and Chairman Mao's Cultural Revolution*, 41–118.

13. Andreas, *Rise of the Red Engineers*, 272–74.

14. On science policies in the Deng era see Tony Saich, *China's Science Policy in the 80s* (Atlantic Highlands: Humanities Press International, 1989), and Miller, *Science and Dissent in Post-Mao China*.

15. See here Yu Guangyuan 于光远, "Chongfen fahui kexue jishu qunzhong tuanti zai sihua jianshezhong de zuoyong 充分发挥科学技术群众团体在四化建设中的作用 (Amply developing the role of the scientific and technological mass organizations in the construction of the Four Modernizations)," *Renmin Ribao*, March 14, 1980.

16. Deng Xiaoping, "Speech at Opening Ceremony of National Science Conference," 13.

17. Margaret Sleeboom-Faulkner, *The Chinese Academy of Science (CASS): Shaping the Reforms, Academia and China* (Leiden: Brill, 2007).

18. "Science and Scientists Revitalized," *Peking Review*, no. 12 (1978): 27–30.

19. Wang Zikun 王梓坤, "Shitan ziran kexue yanjiu de yiban fangfa 试谈自然科学研究的一般方法 (On the general methods of natural science research)," *Hongqi*, no. 2 (1979): 56. In 1985 Wang became the president of Beijing Normal University, and in 1991 was elected as member of the Chinese Academy of Sciences (*Zhongguo kexueyuan*).

20. Li Chang was denounced as an alleged supporter of Liu Shaoqi and purged in 1966. In 1978, he headed a Chinese Academy of Sciences (CAS) delegation to Western Germany and the Netherlands and became the first vice president of the University of Science and Technology.

21. Qian Sanqiang, a nuclear physicist educated in France in the 1930–40s, became professor of nuclear physics at Qinghua University one year before Beijing was occupied by the Communists in 1949 and decided to stay. During the Rectification Campaign, he criticized several scientists as rightist deviationists, and became himself a victim of the Cultural Revolution one decade later, but returned to the CAS in 1973 and headed CAS delegations to Australia, Romania, and Yugoslavia in 1977.

22. The role of Fang Yi, the long-standing member of the CCP Central Committee (1973–87), in connecting Chinese scientists to the outer world was significant. Fang accompanied Deng Xiaoping on the historic visit to the United States in January 1979 during which he traveled to American technological centers in Georgia, Texas, and California.

23. Li Tongyu 李彤宇, "*Ziran bianzhengfa tongxun* chuangkan qianhou 《自然辩证法通讯》创刊前后 (Before and after the foundation of the *Newsletter of Natural Dialectics*)," *Ziran bianzhengfa tongxun* 30, no. 6 (2008): 81–94.

24. See here Shi Xiyuan 石希元, "Lisenke qi ren 李森科其人 (The Person of Lysenko)," *Ziran bianzhengfa tongxun*, no. 1 (1979): 65–81.

25. Wang Zikun, "*Shitan ziran kexue yanjiu de yiban fangfa*."

26. Gu Mainan 顾迈南 and He Huangbiao 何黄彪, "Guowuyuan pizhun liutiao zhengce jiexian 国务院批准六条政策界限 (The State Council approves six policy limits)," *Renmin Ribao*, December 18, 1983. This source has been identified by Sleeboom-Faulkner, *The Chinese Academy of Science*, 69–70.

27. Hu Sheng 胡绳, "Guanyu jiaqiang shehui kexue yanjiu de jige wenti 关于加强社会科学研究的几个问题 (Several issues regarding strengthening social science research)," *Hongqi*, no. 9 (1986): 3–10. In the same year Wu Jianguo 吴建国 made in his 1986 article "Reflections on the Question of Freedom" a distinction between political freedom and philosophical freedom, pointing out that true creative freedom can only exist when the Marxist worldview is ensured. If these two freedoms (the political and the creative) were actually confused, then this could result in the persecution of scientists. See Wu Jianguo 吴建国, "Guanyu ziyou wenti de 'fansi' 关于自由问题的'反思' (Reflections concerning the problem of freedom)," *Hongqi*, no. 17 (1986): 32–38. This article was edited by Hu Qiaomu and reprinted in the national press.

28. Fang Yi 方毅, "Dangqian kexue jishu gongzuo de jige wenti 当前科学技术工作的几个问题 (Several issues regarding current science and technology work)," *Hongqi*, no. 2 (1980): 2–7.

29. "Zongjie jingyan jianchi gaige 总结经验 坚持改革 (Summarizing Experiences—Persisting in Reforms)," *Hongqi*, no. 14 (1982): 2–6.

30. One reason for the necessity to do so was that the number of technical workers was considered too small for raising productivity in the early 1980s. See "Fazhan kexue shiye shi woguo jingji jianshe de zhanlüe zhongdian zhi yi 发展科学事业是我国经济建设的战略重点之一 (Developing science is a strategic focus of our country's economic construction)," *Hongqi*, no. 19 (1982): 15–20. It was considered indispensable to raise the number of technical experts and provide them with machinery and tools, which was to be achieved by science, or *kexue*.

31. Fan Dainian and Robert S. Cohen, eds., *Chinese Studies in the History and Philosophy of Science and Technology* (Dordrecht: Kluwer Academic Publishers, 1996); Miller, *Science and Dissent in Post-Mao China*; Orleans, *Science in Contemporary China*; Saich, *China's Science Policy in the 80s*; Wilson and Keeley, *China, the Next Science Superpower?*

32. And even the Fukushima disaster did not cause a reconsideration of nuclear power among Chinese government officials, see Matten, "Coping with Invisible Threats."

33. See here the compelling and also somewhat depressing view in Anthony Giddens, *The Consequences of Modernity* (Oxford: Polity Press, 1990).

34. See the two texts by Deng Xiaoping 邓小平, "Zunzhong zhishi, zunzhong rencai 尊重知识, 尊重人才 (Respecting Knowledge, Respecting Talents)," *Deng Xiaoping wenxuan*, 2 (1993): 40–41; and "Kexue jishu shi diyi shengchanli 科学技术是第一生产力 (Science and Technology is the primary productive force)," *Deng Xiaoping wenxuan* 3 (1993): 274–76.

35. For the chicken blood therapy see Joel Martinsen, "Injecting Chicken Blood," accessed January 3, 2021, http://www.chinaheritagequarterly.org/articles.php?searchterm=025_blood.inc&issue=025.

36. See exemplarily the discussions in Wen-yuan Lin and John Law, "We Have Never Been Latecomers!? Making Knowledge Spaces for East Asian Technosocial Practices," *East Asian Science, Technology and Society: An International Journal*, no. 9 (2015): 117–26; Law and Wen-yuan Lin, "Provincializing STS"; Palmer, *Qigong Fever*; Fa-ti Fan, "East Asian STS: Fox or Hedgehog?"; Fa-ti Fan, "Science, State, and Citizens." On a specific case of how to rescue traditional practices in the modern age see Zhu Shine 祝世讷, *Zhongyi wenhua de fuxing* 中医文化的复兴 (Rejuvenation of Chinese medicine culture) (Nanjing: Nanjing chubanshe, 2013).

37. On the discussions of these special abilities in 1980s Chinese society and their dispute with natural sciences see the introduction in Qian Xuesen 钱学森, *Lun renti kexue* 论人体科学 (On Somatic Science) (Beijing: Renmin junyi chubanshe, 1988). On a Marxist critique of the concept of special abilities see the discussion in Yu Guangyuan 于光远, *Zhongguo de kexue jishu zhexue: ziran bianzhengfa* 中国的科学技术哲学: 自然辩证法 (Philosophy of science and technology in China: natural dialectics) (Beijing: Kexue chubanshe, 2013), 133–46, 171–92.

38. In 1995 a nation-wide campaign against superstition was started by the State Council's circular on "scientific popularization work," see He Zuoxiu 何祚庥, *Wo shi He Zuoxiu* 我是何祚庥 (I am He Zuoxiu) (Beijing: Zhongguo shidai jingji chu-

banshe, 2002), 12. Two weeks later, the *People's Daily* declared that fighting against superstition became necessary again, because mantic techniques were popular in rural areas and threatened to spread to the cities. An article in the *People's Daily* said these techniques scientific was deceiving the population and counterproductive to the nation's modernization, see Zhao Zhenyu 赵振宇, "Kexue, burong mixin dianwu 科学，不容迷信玷污 (Science cannot tolerate the stains of superstition)," *Renmin Ribao*, April 26, 1995.

39. For a short biography of He Zuoxiu see Miller, *Science and Dissent in Post-Mao China*.

40. To solve this problem Fang Zhouzi proposes a differentiation between Chinese medical theory and Chinese medicine. The former can hardly be reconciled with Western medical theory (even though some were fantasizing that it could be integrated into the universal science system). See Fang Zhouzi 方舟子, "Kexue de kandai Zhong yiyao 科学地看待中医药 (Looking at Chinese Medicine Scientifically)," *Diyi caijing ribao*, March 1, 2006. The turn to culture can be observed in the changing content of two journals, *Medicine and Philosophy* (*Yixue yu zhexue* 医学与哲学) and *History of Chinese Medicine* (*Zhonghua yishi* 中华医史). In the case of the former, founded in the year 1980, culture is introduced as an epistemological principle that defines medical science in its volumes published during the 1990s. Having replaced ideology, culture not only allows a far more flexible definition but also makes it more attractive for a populace that by now has subscribed to the equation of culture to patriotic pride. It is important to note in this context that the cultural pride certainly helped propagating traditional knowledge, such as the pride on Chinese medical therapies. Its defenders argue that these therapies cannot leave their cultural context behind; in other words, its particularity in terms of medicine is a reflection of its cultural tradition. See Men Jiuzhang 门九章 and Chang Cunku 常存库, "Zhong-Xiyi jiehe de wenhua sikao—Zhong-Xi jiehe de biranxing lunzheng 中西医结合的文化思考—中西医结合的必然性论证 (Cultural Reflections on the Combination of Chinese and Western Medicine—The necessary proof that Chinese and Western Medicine will merge)," *Yixue yu zhexue*, no. 5 (1998): 250–51.

41. Peter Burke, *What Is the History of Knowledge?* (Cambridge: Polity Press, 2016).

42. Lan Liji 蓝礼吉, "Houxiandai yixue wenhua yu renwen shehui yixue de chengzhang 后现代医学文化与人文社会医学的成长 (Postmodern Medical Culture and the Growth of Medical Science in the Humanities and Social Sciences)," *Yixue yu zhexue*, no. 5 (1998): 255–56. See also Qiu Hongzhong 邱鸿钟, *Yixue yu renlei wenhua: yixue wenhua shehuixue yinlun* 医学与人类文化: 医学文化社会学引论 (Medicine and Human Culture: A Sociological Introduction to Medical Culture) (Changsha: Hunan kexue jishu chubanshe, 1993).

43. See here Nie Jingbao 聂精葆, "Kexue zhuyi longzhaoxia de 20 shiji Zhongyi—jianlun Zhongyi shifou shi kexue 科学主义笼罩下的20世纪中医－兼论中医是否是科学 (Twentieth Century Chinese Medicine under Scientism—Also on whether Chinese Medicine Is Scientific)," *Yixue yu zhexue*, no. 2 (1995): 62–66.

44. The new category was named *Interdisciplinary works and borderline sciences (including scientific anomalies and discussions of "pseudoscience")*. See here the

bibliography published on http://isisbibliography.org/archive/, last accessed July 1, 2021. For possible consequences of such a move see Michael Hagner, "Bye-bye Science, Welcome Pseudoscience? Reflexionen über einen beschädigten Status," in *Pseudowissenschaft*, ed. Dirk Rupnow, Veronika Lipphardt, Jens Thiel, and Christina Wessely (Frankfurt: Suhrkamp, 2008), 21–50.

45. Peter Dear, "What Is the History of Science the History *Of*? Early Modern Roots of the Ideology of Modern Science," *Isis* 96 (2005): 390–406.

46. Elshakry describes how the Western differentiation between technical knowledge and natural philosophical knowledge led to the rejection of Eastern contributions to science because these were not based on mathematical or empirical methods. In her view, this also explains why Needham concentrates in his opus magnum on the applied sciences without sufficiently considering their natural philosophical framework. See Marwa Elshakry, "When Science Became Western: Historiographical Reflections," *Isis* 101, no. 1 (March 2010): 98–109, here 99.

47. Characteristic for such a kind of thinking are questions such as "Why did China not develop modern science (as did Europe)?," cf. here Lin, Justin Yifu, "The Needham Puzzle: Why the Industrial Revolution Did Not Originate in China," *Economic Development and Cultural Change* 43, no. 2 (1995): 269–92. This universalizing notion was caused by the expansion of Western powers when Europe's military and technological supremacy was seen as evidence of the efficacy of Western science. The appropriation of Western science by the colonized countries was either radical due to calls for total Westernization, or partial by syncretism of foreign and domestic forms of sciences. Cf. here Benjamin Elman, "Universal Science versus Chinese Science: The Changing Identity of Natural Studies in China, 1850–1930," *Historiography East and West*, no. 1 (2003): 70–116.

48. Burke, *What Is the History of Knowledge?*

49. It was preceded by the 2002 National Law on the Dissemination of Science and Technology (*Zhonghua renmin gongheguo kexue jishu pujifa* 中华人民共和国科学技术普及法) that set up the National Day of Science Popularization (*Quanguo kepuri* 全国科普日).

Bibliography

Agarkow, A., E. Leonenko, and G. Müller. "Shaft Drilling in the U.S.S.R.: History and Recent Experience." In *Shaft Engineering*, edited by Institute of Mining and Metallurgy, 1–32. London: Taylor & Francis e-Library, 2005/1989.

Amelung, Iwo, Matthias Koch, Joachim Kurtz, Eun-Jeung Lee, and Sven Saaler, eds. *Selbstbehauptungsdiskurse in Asien: China—Japan—Korea*. München: Iudicium, 2003.

American Insect Control Delegation. *Insect Control in the People's Republic of China: A Trip Report of the American Insect Control Delegation, Submitted to the Committee on Scholarly Communication with the People's Republic of China*. Washington: National Academy of Sciences, 1977.

American Plant Studies Delegation. *Plant Studies in the People's Republic of China: A Trip Report of the American Plant Studies Delegation*. Washington, D.C.: National Academy of Sciences, 1975.

Anderson, Warwick. "Postcolonial Specters of STS." *East Asian Science, Technology and Society* 11, no. 2 (June 2017): 229–33.

Andreas, Joel. *Rise of the Red Engineers: The Cultural Revolution and the Origin of China's New Class*. Stanford: Stanford University Press, 2009.

Anhui sheng nongyeting xumu shouyichu 安徽省农业厅畜牧兽医处. *Anhui zhongshouyi yaofang huibian* 安徽中兽医药方汇编 (Collected Recipes of Chinese Veterinary Medicine in Anhui). Hefei: Anhui renmin chubanshe, 1958.

Ariès, Philippe. *Centuries of Childhood: A Social History of Family Life*. New York: Alfred A. Knopf, 1962.

Baark, Erik. "The Structure of Technological Information Dissemination in China: Publication of Scientific and Technological Manuals, 1970–77." *China Quarterly*, no. 83 (September 1980): 510–34.

Barnouin, Barbara, and Changgen Yu. *Zhou Enlai: A Political Life*. Hong Kong: Chinese University Press, 2006.

Bernstein, Thomas P., and Hua-Yu Li, eds. *China Learns from the Soviet Union: 1949–Present*. Lanham: Lexington Books, 2010.

Bi, Yuan 毕苑. *Jianzao changshi: jiaokeshu yu jindai zhongguo wenhua zhuanxing* 建造常识： 教科书与近代中国文化转型 (Establishing Common Knowledge: Textbook and the Transformation of Modern Chinese Culture). Fuzhou: Fujian jiaoyu chubanshe, 2010.

Bo, Yibo 薄一波. *Ruogan zhongda juece yu shijian de huigu (shangjuan)* 若干重大决策与事件的回顾 (上卷) (Reviewing a number of significant strategic decisions and events [vol. 1]). Beijing: Renmin chubanshe, 1991.

Bowers, John Z., J. William Hess, and Nathan Sivin, eds. *Science and Medicine in Twentieth-Century China: Research and Education.* Ann Arbor: Center for Chinese Studies, University of Michigan, 1988.

Bowie, Robert R., and John K. Fairbank, eds. *Communist China 1955–1959: Policy Documents with Analysis.* Cambridge: Harvard University Press, 1965.

Bray, Francesca. "Chinese Literati and the Transmission of Technological Knowledge: The Case of Agriculture." In *Cultures of Knowledge: Technology in Chinese History*, edited by Dagmar Schäfer, 299–326. Leiden: Brill, 2012.

Bray, Francesca. *Technology and Gender: Fabrics of Power in Late Imperial China.* Berkeley: University of California Press, 1997.

Broadbent, Kieran P. *Dissemination of Scientific Information in the People's Republic of China.* Ottawa: International Development Research Centre, 1980.

Brock, Darryl E. "The People's Landscape: Mr. Science and the Mass Line." In *Mr. Science and Chairman Mao's Cultural Revolution*, edited by Chunjuan N. Wei and Darryl E. Brock, 41–118. Lanham: Lexington Books, 2013.

Brown, Jeremy, and Matthew D. Johnson, eds. *Maoism at the Grassroots: Everyday Life in China's Era of High Socialism.* Cambridge: Harvard University Press, 2015.

Brugger, William. *Democracy and Organisation in the Chinese Industrial Enterprise (1948–1953).* Cambridge: Cambridge University Press, 1976.

Bruun, Ole. *Fengshui in China: Geomantic Divination between State Orthodoxy and Popular Religion.* Honolulu: University of Hawai'i Press, 2003.

Bunge, Mario. "Demarcating Science from Pseudoscience." *Fundamenta scientiae*, no. 2 (1982): 369–88.

Burke, Peter. *What Is the History of Knowledge?* Cambridge: Polity Press, 2016.

Byrd, William A., ed. *Chinese Industrial Firms under Reform.* Oxford: Oxford University Press, 1992.

Cao, Guangzhe 曹光哲, Qi Pengfei 齐鹏飞, and Wang Jin 王进, eds. *Zhou Enlai dacidian* 周恩来大辞典 (Great Dictionary on Zhou Enlai). Guilin: Guangxi chubanshe, 1997.

Chakrabarty, Dipesh. *Provincializing Europe: Postcolonial Thought and Historical Difference.* Princeton, NJ: Princeton University Press, 2008.

Chaliaofu 察廖夫 (G. Tsarev). *Gongchang dang zuzhi zenyang lingdao shehuizhuyi jingsai* 工厂党组织怎样领导社会主义竞赛 (How does the party in factory lead socialist competitions?), translated by Chen Dawei 陈大维. Beijing: Shidai chubanshe, 1951.

Chan, Alfred L. *Mao's Crusade: Politics and Policy Implementation in China's Great Leap Forward.* Oxford: Oxford University Press, 2001.

Chaosheng 潮生. "Shishi qiushi 实事求是 (Seeking Truth from Facts)." *Tongwen yuekan* 通问月刊, no. 4 (1931): 1–2.

Chen, Hongjie 陈洪杰. "Zhongguo jindai kepu jiaoyu: shetuan, changguan he jishu 中国近代科普教育：社团，场馆和技术 (Modern Science Education in China: Societies, Spaces, and Technologies)." MA thesis, East China Normal University, 2006.

Chen, Huan 陈焕. "Yong niushi wei zhu de banfa hao 用牛屎喂猪的办法好 (The Method of Using Cow Dung to Feed Pigs Is Good)." In *Fazhan shengxu shengchan de jingyan* 发展生畜生产的经验 (*Nongye shengchan jingyan congshu* 农业生产经验丛书), ed. Zhong-Gong Guangdong sheng weiyuanhui bangongting 中共广东省委员会办公厅, 36–40. Guangzhou: Guangdong renmin chubanshe, 1958.

Chen, Sihe 陈思和. "Ruhe dangjia? Zenyang zuozhu? Chongdu Lu Mei zhibi de huaju 'hongqi ge' 如何当家？怎样做主？重读鲁煤执笔的话剧《红旗歌》(How to Become the Master? Rereading the Spoken Drama *Red Flag Song* Written by Lu Mei)." *Zhongguo xiandai wenxue yanjiu congkan* 中国现代文学研究丛刊, no. 4 (2011): 31–42.

Chu, Pingyi. "Narrating a History for China's Medical Past: Christianity, Natural Philosophy and History in Wang Honghan's *Gujin yishi* 古今醫史 (History of Medicine Past and Present)." *EASTM*, no. 28 (2008): 14–35.

Clark, M. Gardner. *Development of China's Steel Industry and Soviet Technical Aid.* Ithaca: Cornell University Press, 1973.

Cooper, William, and Nathan Sivin. "Man As a Medicine: Pharmacological and Ritual Aspects of Traditional Therapy Using Drugs Derived from the Human Body." In *Chinese Science: Exploration of an Ancient Tradition*, edited by Shigeru Nakayama and Nathan Sivin, 203–72. Cambridge: MIT Press, 1973.

Crow, James. "Genetics in Postwar China." In *Science and Medicine in Twentieth-Century China*, edited by Bowers, Hess, and Sivin, 155–70. Ann Arbor: Center for Chinese Studies, University of Michigan, 1988.

Croizier, Ralph C. *Traditional Medicine in Modern China: Science, Nationalism, and the Tensions of Cultural Change.* Cambridge: Harvard University Press, 1968.

Dai, Maolin 戴茂林. "Angang xianfa yanjiu 鞍钢宪法研究 (A Study of the AnSteel Charter)." *Zhonggong dangshi yanjiu*, no. 6 (1999): 38–43.

Daston, Lorraine, and Peter Galison. *Objectivity.* New York: Zone Books, 2010.

Dear, Peter. "What Is the History of Science the History *Of*? Early Modern Roots of the Ideology of Modern Science." *Isis* 96 (2005): 390–406.

DeMare, Brian James. "Casting (Off) Their Stinking Airs: Chinese Intellectuals and Land Reform, 1946–52." *The China Journal*, no. 67 (January 2012): 109–29.

Deng, Xiaoping. "Speech at Opening Ceremony of National Science Conference." *Peking Review*, no. 12 (1978): 9–18.

Deng Xiaoping Selected Works. Beijing: Foreign Languages Press, 1994.

Deng Xiaoping wenxuan 邓小平文选. Beijing: Renmin chubanshe, 1993.

Dianying puji 电影普及 (*Popularization of Films*). 1981–87.

Dianying yishu 电影艺术 (*Film Art*). 1956–94.

Dongbei nongye 东北农业 (*Agriculture in the Northeast*). 1950.

Du, Nianxing 杜念兴. *Shouyixue dayi* 兽医学大意 (An Overview of Veterinary Medicine). Nanjing: Xumu shouyi tushu chubanshe, 1957.

Du, Runsheng. "Great Progress Made in the Natural Sciences in China During the Last Decade." *The Science News-Letter* 78, no. 24 (1960): 377–92, originally appearing in *Scientia Sinica* VIII, no. 11 (1959).

Eisenman, Joshua. *Red China's Green Revolution: Technological Innovation, Institutional Change, and Economic Development Under the Commune*. New York: Columbia University Press, 2018.

Elman, Benjamin A. *A Cultural History of Modern Science in China*. Cambridge: Harvard University Press, 2006.

Elman, Benjamin A. *On Their Own Terms: Science in China, 1550–1900*. Cambridge: Harvard University Press, 2005.

Elman, Benjamin. "Universal Science versus Chinese Science: The Changing Identity of Natural Studies in China, 1850–1930." *Historiography East and West*, no. 1 (2003): 70–116.

Elshakry, Marwa. "When Science Became Western: Historiographical Reflections." *Isis* 101, no. 1 (March 2010): 98–109.

Elvin, Mark. "The High-Level Equilibrium Trap: The Causes of the Decline of Invention in the Traditional Chinese Textile Industries." In *Economic Organization in Chinese Society*, edited by W. E. Willmott, 137–72. Stanford: Stanford University Press, 1972.

Fan, Dainian, and Robert S. Cohen, eds. *Chinese Studies in the History and Philosophy of Science and Technology*. Dordrecht: Kluwer Academic Publishers, 1996.

Fan, Fa-ti. "East Asian STS: Fox or Hedgehog?" *East Asian Science, Technology and Society* 1, no. 2 (2007): 243–47.

Fan, Fa-ti. "Science, State, and Citizens: Notes from Another Shore." *Osiris* 27, no. 1, Clio Meets Science: The Challenges of History (2012): 227–49.

Fan, Ka-wai. "Film Propaganda and the Anti-schistosomiasis Campaign in Communist China." *Sungkyun Journal of East Asian Studies* 12, no. 1 (April 2012): 1–17.

Fang, Xiaoping. *Barefoot Doctors and Western Medicine in China*. Rochester: University of Rochester Press, 2012.

Farquhar, Judith. *Knowing Practice: The Clinical Encounter of Chinese Medicine*. San Francisco: Westview Press, 1994.

Feuchtwang, Stephan. "The Problem of 'Superstition' in the People's Republic of China." In *Religion and political power*, edited by Gustavo Benavides and Martin W. Daly, 43–68. Albany: State University of New York Press, 1989.

First Five-Year Plan for Development of the National Economy of the People's Republic of China in 1953–57. Beijing: Foreign Languages University Press, 1956.

Frank, Andre Gunder. *ReORIENT. Global Economy in the Asian Age*. Berkeley and Los Angeles, California: University of California Press, 1998.

Fröhlich, Thomas. *Staatsdenken im China der Republikzeit (1912–1949): Die Instrumentalisierung philosophischer Ideen bei chinesischen Intellektuellen*. Frankfurt: Campus Verlag, 2000.

Fujian sheng nongye ting ed. 福建省农业厅. *Yangzhu shouce* 养猪手册 (Handbook on keeping pigs). Fuzhou: Fujian renmin chubanshe, 1957.

Fujian sheng nongye ting ed. 福建省农业厅. *Yangzhu siliao duo de hen yangzhu siliao duo de hen* 养猪饲料多得很 (Raising pigs with fodder on a large scale). Fuzhou: Fujian renmin chubanshe, 1959.

Gailiang nongju jieshao 改良农具介绍 (Introducing Improvements of Farming Tools), 1956.

Galison, Peter, and David J. Stump, eds. *The Disunity of Science: Boundaries, Contexts, and Power.* Stanford: Stanford University Press, 1996.

Gangren tiema kua xincheng 钢人铁马跨新程 (Steel Man Riding Iron Horse on the New Road). Beijing: Yejin gongye chubanshe, 1974.

Gangtie 钢铁 (Steel and Iron). 1958–1961.

Gao, Hua 高華. "Angang xianfa de lishi zhenshi yu 'zhengzhi zhengquexing' 鞍鋼憲法的歷史真實與「政治正確性」 (The Historical Authenticity of the AnSteel Charter and 'Political Correctness')." *Ershiyi shiji*, no. 58 (2000): 62–69.

Gaojinnawa 高金娜娃 (D. Goginawa). *Qiye zhong de dang guanli* 企业中的党管理 (The party's administration in enterprises), translated by Cao Ying 草婴. Shanghai: Shidai chubanshe, 1950.

Ghosh, Arunabh. *Making it Count: Statistics and Statecraft in the Early People's Republic of China.* Princeton, NJ: Princeton University Press, 2020.

Giddens, Anthony. *The Consequences of Modernity.* Oxford: Polity Press, 1990.

Goikhman, Izabella. "Soviet-Chinese Academic Interactions in the 1950s: Questioning the 'Impact-Response' Approach." In *China Learns from the Soviet Union, 1949–Present,* edited by Thomas P. Bernstein and Hua-yu Li, 275–302. New York: Rowman and Littlefield.

Gongye de jichu—Gangtie gongye 工业的基础—钢铁工业 (Foundation of Industry—the Iron and Steel Industry). Beijing: Zhonghua quanguan kexue jishu puji chubanshe, 1954.

Grieder, Jerome. *Hu Shih and the Chinese Renaissance: Liberalism in the Chinese Revolution, 1917–1937.* Cambridge: Harvard University Press, 1970.

Gross, Miriam. *Farewell to the God of Plague: Chairman Mao's Campaign to Deworm China.* Berkeley: University of California Press, 2016.

Guowuyuan guanyu minjian shouyi gongzuo de jueding 国务院关于民间兽医工作的决定 (The State Council's Decision on Folk Veterinary Work). In "Hunan sheng renmin weiyuanhui zhuanfa guowuyuan guanyu minjian shouyi gongzuo de jueding 湖南省人民委员会转发国务院关于民间兽医工作的决定." *Hunan zhengbao*, no. 9 (1963): 4–6.

Hacking, Ian. *Representing and Intervening: Introductory Topics in the Philosophy of Natural Science.* Cambridge: Cambridge University Press, 1983.

Hagner, Michael. "Bye-bye Science, Welcome Pseudoscience? Reflexionen über einen beschädigten Status." In *Pseudowissenschaft*, edited by Dirk Rupnow, Veronika Lipphardt, Jens Thiel, and Christina Wessely, 21–50. Frankfurt: Suhrkamp, 2008.

Hao, Shijian 郝石坚. *Xiandai qiumo zhutie* 现代球墨铸铁 (Modern nodular cast iron). Beijing: Meitan gongye chubanshe, 1989.

Harding, Sandra. *Sciences from Below: Feminisms, Postcolonialities, and Modernities.* Durham: Duke University Press, 2008.

He, Zuoxiu 何祚庥. *Wo shi He Zuoxiu* 我是何祚庥 (I am He Zuoxiu). Beijing: Zhongguo shidai jingji chubanshe, 2002.

Hebei zhongshouyi xuexiao 河北中兽医学校, Beijing budui houqinbu weishengbu 北京部队后勤部卫生部, Dongbei nongken daxue 东北农垦大学, Zhou xian shouyiyuan 涿县兽医院. *Zhongshouyi shouce* 中兽医手册 (Manual of Chinese Veterinary Medicine). Beijing: Nongye chubanshe, 1975.

Heilongjiang sheng nongyeting xumu shouyichu eds. 黑龙江省农业厅畜牧兽医处编. *Minjian shouyi yanfang* 民间兽医验方 (Tested Drug Formulae of Folk Veterinary Medicine). Harbin: Heilongjiang renmin chubanshe, 1958.

Hinton, William. "The Importance of Land Reform in the Reconstruction of China." *Monthly Review* 50, no. 3 (1998): 147–60.

Hinton, William. *Shenfan*. New York: Random House, 1983.

Ho, Denise Y., and Li Jie. "From Landlord Manor to Red Memorabilia: Reincarnations of a Chinese Museum Town." *Modern China* 42, no. 1 (2016): 3–37.

Ho, Denise Y. *Curating Revolution: Politics on Display in Mao's China*. Cambridge: Cambridge University Press, 2017.

Hong, Xiangsheng 洪祥生. "Ba zhonghua minzu zaojiu wei yige kexue de minzu: Xi Tao Xingzhi de 'kexue xiajia' 把中华民族造就为一个科学的民族：析陶行知的'科学下嫁' (Making the Chinese Nation into a Scientific Nation: An Analysis of Tao Xingzhi's 'Marrying Science Down' Movement)." *Anhui jiaoyu xueyuan xuebao*, no. 1 (1988): 79–81.

Hongqi 红旗 (*Red Flag*). 1958–88.

Hsu, Elisabeth. "The History of Chinese Medicine in the People's Republic of China and Its Globalization." *East Asian Science, Technology and Society* 2, no. 4 (2008): 465–84.

Hsu, Elisabeth. *The Transmission of Chinese Medicine*. Cambridge: Cambridge University Press, 1999.

Hu, Danian. *China and Albert Einstein: The Reception of the Physicist and His Theory in China, 1917–1979*. Cambridge: Harvard University Press, 2005.

Hu, Shi 胡适. *Chabuduo xiansheng zhuan* 差不多先生传 (Biography of Mr. More-or-Less). Shanghai: Shaonian ertong chubanshe, 1924.

Hua, Luogeng 华罗庚. *Gei qingnian shuxuejia* 给青年数学家 (To Young Mathematicians). Beijing: Zhongguo qingnian chubanshe, 1956.

Hua Luogeng 华罗庚, Tongchou fangfa pinghua ji buchong 统筹方法平话及补充 (Popular stories and supplements of CPM methods). Bejing: Zhongguo gongye chubanshe, 1965.

Huadong junzheng weiyuanhui tudi gaige weiyuanhui 华东军政委员会土地改革委员会, ed. *Shanghai shijiaoqu Sunan xingzhengqu—tudi gaige huaji* 上海市郊区苏南行政区—土地改革画集 (Sunan District in Shanghai Suburb: Pictorial on Land Reform). Huadong junzheng weiyuanhui tudi gaige weiyuanhui, 1952.

Huang, Yong 黄勇. "Youpai wenxuezhong de zirankexuejia 右派文學中的自然科學家 (Natural Scientists in the Rightist Literature)." *Ershiyi shiji*, no. 110 (2008): 79–88.

Institute of Literature of the Chinese Academy of Social Sciences. *Stories about Not Being Afraid of Ghosts*, trans. Yang Hsien-yi and Gladys Yang. Peking: Foreign Languages Press, 1961.

Isett, Christopher M. *State, Peasant, and Merchant on the Manchurian Frontier, 1644–1862*. Stanford: Stanford University Press, 2007.

Ishii, Yumi 石井弓. *Kioku toshite no Nitchū sensō: intabyū ni yoru tasha rikai no kanōsei* 記憶としての日中戦争：インタビューによる他者理解の可能性 (The Japanese-Chinese War as memory: possibility of understanding others by interview). Tōkyō: Kenbun shuppan, 2013.

J. V. Stalin Works (13 vols.). Moscow: Foreign Languages Publishing House, 1954.

Jersild, Austin. *The Sino-Soviet Alliance: An International History (New Cold War History)*. North Carolina: University of North Carolina Press, 2014.

Jiangsu sheng nonglinting nongkenju 江苏省农林厅农垦局 and Jiangsu sheng shougongye guanliju 江苏省手工业管理局. *Nongju gaige congshu: Qunzhong chuangzao de zhonggeng chucao nongju* 农具改革丛书：群众创造的种耕除草农具 (Book Series on Agricultural Tool Improvement: Plowing and weeding tools created by the masses). Nanjing: Jiangsu renmin chubanshe, 1958.

Jiangsu sheng nongye kexue yanjiusuo geming weiyuanhui 江苏省农业科学研究所革命委员会. *Zhongshouyi zhenliao xuanbian* 中兽医诊疗选编 (Selected Cases of Chinese Veterinary Medicine Diagnosis and Treatment). Nanjing: Jiangsu sheng geming weiyuanhui, 1970.

Jiangsu sheng shuiliting 江苏省水利厅, ed. *Qunzhong chuangzao de watu, hangtu gongju* 群众创造的挖土、夯土工具 (Earth Digging and Ramming Tools Created by the Masses). Nanjing: Jiangsu renmin chubanshe, 1958.

Jianguo yilai Mao Zedong wengao 建国以来毛泽东文稿 (Mao Zedong's Manuscripts after 1949, 13 vols.). Beijing: Zhongyang wenxian chubanshe, 1987–98.

Jiangxi sheng nongye kexue yanjiusuo zhongshouyi yanjiushi 江西省农业科学研究所中兽医研究室. *Zhu chuanqibing zhenjiu liaofa* 猪喘气病针灸疗法 (Acupuncture and Moxibustion to Treat Swine Asthma). Nanjing: Jiangsu renmin chubanshe, 1959.

Jiangxi sheng nongyeting zhongshouyi shiyansuo 江西省农业厅中兽医实验所. *Shiyong shouyi zhenjiuxue* 实用兽医针灸学 (Practical Acupuncture and Moxibustion in Veterinary Medicine). Nanchang: Jiangxi renmin chubanshe, 1958.

Jiangxi sheng nongyeting 江西省农业厅, ed. *Gailiang nongju de zhizao yu shiyong* 改良农具的制造与使用 (Construction and Use of Improved Farming Tools). Nanchang: Jiangxi renmin chubanshe, 1955.

Jiangxi sheng nongyeting 江西农业厅, and Shouyi fangzhizhan 兽医防治站, eds. *Tubing fangzhifa* 兔病防治法 (Prevention and treatment of rabbit diseases). Beijing. Nongye chubanshe, 1959.

Jiangxi sheng xumu shouyi gongzuozhe xiehui 江西省畜牧兽医工作者协会, and Jiangxi sheng nongyeting zhongshouyi shiyansuo 江西省农业厅中兽医实验所. *Jiangxi minjian shouyi zhenliao ji chufang huibian* 江西民间兽医诊疗及处方汇编 (Compilation of folk veterinary diagnosis and formulae in Jiangxi). Nanchang: Jiangxi renmin chubanshe, 1956.

Jingji ziliao bianji weiyuanhui 经济资料编辑委员会. *Zenyang tigao tuolaji de gongzuo xiaolü* 怎样提高拖拉机的工作效率 (How to increase the working efficiency of tractors). Beijing: Caizheng jingji chubanshe, 1956.

Jixie gongren 机械工人 (*Industry Workers*). 1950–56.

Johnson, Matthew D. "The Science Education Film: Cinematizing Technocracy and Internationalizing Development." *Journal of Chinese Cinemas* special issue, "The Missing Period of PRC Cinema" 5, no. 1 (2011): 31–53.

Käding, Edda. *Engagement und Verantwortung. Hans Stubbe, Genetiker und Züchtungsforscher. Eine Biographie.* Müncheberg, 1999.

Kaple, Deborah A. *Dream of a Red Factory: The Legacy of High Stalinism in China.* Oxford: Oxford University Press, 1994.

Kaple, Deborah A. "Soviet Advisors in China in the 1950s." In *Brothers in Arms: The Rise and Fall of the Sino-Soviet Alliance, 1945–1963*, edited by Odd Arne Westad, 117–40. Washington: Woodrow Wilson Center Press, 1998.

Kelimannuofu 克里曼诺夫 (A. Klimanov). *Qiye zhong de dang gongzuo* 企业中的党工作 (The party's work in enterprises), translated by Cao Ying 草婴. Shanghai: Shidai chubanshe, 1950.

Kexue dazhong 科学大众 (*Popular Science*). 1946–66.

Kexue huabao 科学画报 (*Science Pictorial*). 1933–66.

Kexue puji gongzuo 科学普及工作 (*Science Dissemination Work*). 1951–58.

Kexue puji tongxun 科学普及通讯 (*Newsletter of Science Dissemination*). 1950–51.

Kexue puji ziliao 科学普及资料 (*Materials on Dissemination of Science*). 1972–74.

Khlevniuk, Oleg V. *In Stalin's Shadow: The Career of "Sergo" Ordzhonikidze.* Armonk: M. E. Sharpe, 1995.

Kirby, William C. "Engineering China: Birth of the Developmental State, 1928–1937." In *Becoming Chinese: Passages to Modernity and Beyond*, edited by Wen-hsin Yeh, 137–60. Berkeley: University of California Press, 2000.

Knorr-Cetina, Karin. "The Care of the Self and Blind Variation: The Disunity of Two Leading Sciences." In *The Disunity of Science: Boundaries, Contexts, and Power*, edited by Peter Galison and David Stump, 287–310. Stanford: Stanford University Press, 1996.

Kotkin, Stephen. *Magnetic Mountain: Stalinism as a Civilization.* Berkeley: University of California Press, 1995.

Kuromiya, Hiroaki. "Edinonachalie and the Soviet Industrial Manager, 1928–1937." *Soviet Studies* 36, no. 2 (1984): 185–204.

Kwok, Daniel W. Y. *Scientism in Chinese Thought 1900–1950.* London: Yale University Press, 1965.

Lam, Tong. *A Passion for Facts: Social Surveys and the Construction of the Chinese Nation-State, 1900–1949.* Berkeley: University of California Press, 2011.

Latour, Bruno. *Politics of Nature: How to Bring the Sciences into Democracy.* Translated by Catherine Porter. Cambridge: Harvard University Press, 2004.

Latour, Bruno. *We Have Never Been Modern.* Translated by Catherine Porter. Cambridge: Harvard University Press, 1993.

Lavine, Steven D., and Ivan Karp. "Introduction: Museums and Multiculturalism." In *Exhibiting Cultures: The Poetics and Politics of Museum Display*, edited by Steven D. Lavine and Ivan Karp, 1–9. Washington: Smithsonian Institution Press, 1991.

Law, John, and Wen-yuan Lin. "Provincializing STS: Postcoloniality, Symmetry, and Method." *East Asian Science, Technology and Society: An International Journal*, no. 11 (2017): 211–27.

Lee, Hong Yung. *From Revolutionary Cadres to Party Technocrats in Socialist China*. Berkeley: University of California Press, 1991.

Lei, Sean Hsiang-lin. "How Did the Chinese Medicine Become Experiential? The Political Epistemology of Jingyan." *positions* 10, no. 2 (2002): 333–64.

Lei, Sean Hsiang-lin. *Neither Donkey nor Horse: Medicine in the Struggle over China's Modernity*. Chicago: University of Chicago Press, 2014.

Li, Huazhong 李华忠, and Zhang Yu 张羽. *Angang sishi nian* 鞍钢四十年 (Forty Years of the AnSteel). Shenyang: Liaoning renmin chubanshe, 1989.

Li, Peishan, Meng Qinzhe, Huang Qinghe, and Huang Shu-e. "The Qingdao Conference of 1956 on Genetics: The Historical Background and Fundamental Experiences." In *Chinese Studies in the History and Philosophy of Science and Technology*, edited by Robert S. Choen, and Dainian Fan, 41–54. Boston: Kluwer Academic Publishers, 1996.

Lin, Justin Yifu. "The Needham Puzzle: Why the Industrial Revolution Did Not Originate in China." *Economic Development and Cultural Change* 43, no. 2 (1995): 269–92.

Lin, Liang 林亮. "Shijian fencuile zichanjieji de weikexue 实践粉碎了资产阶级的伪科学 (Pratice smashes the bourgeois pseudo-science)." *Chuangzao*, no. 2 (1958): 31–34

Lin, Shan 林山. "Jixu guzu ganjin, shixian jinnian nongye kexue gengda de yuejin! 继续鼓足干劲，实现今年农业科学更大的跃进! (Continue to do one's best to realize this year's even bigger leap forward in agricultural science)." *Guangxi nongye kexue* 12 (1959): 1–4, 29.

Lin, Wen-yuan, and John Law. "We Have Never Been Latecomers!? Making Knowledge Spaces for East Asian Technosocial Practices." *East Asian Science, Technology and Society: An International Journal*, no. 9 (2015): 117–26.

Liu, Bangfan 刘邦凡. "Hua Luogeng kepu jiaoyu sixiang yanjiu 华罗庚科普教育思想研究 (A study on Hua Luogeng's educational thoughts on science popularization)." *Qinghai minzu shizhuan xuebao*, no. 1 (1999): 1–5.

Liu, Lydia H. *Translingual Practice: Literature, National Culture, and Translated Modernity—China, 1900–1937*. Stanford: Stanford University Press, 1995.

Liu, Ruisan 刘瑞三 and Shen Yancheng 沈延成. *Jiachu siyang* 家畜饲养 (Breeding livestock). Shanghai: Shanghai kexue puji, 1958.

Liu, Zijiu 刘子久. *Laodong jingsai jianghua* 劳动竞赛讲话 (Speeches on Labor Competition). Beijing: Gongren chubanshe, 1954.

Lu, Siman 陆思曼. *Jiachu siliao peizhi jiben zhishi jianghua* 家畜饲料配制基本知识讲话 (Addressing basic knowledge of food preparation for livestock). Shanghai: Kexue jishu chubanshe, 1957.

Luk, Christine Yi Lai. *A History of Biophysics in Contemporary China*. Cham: Springer, 2015.

Lun woguo de kexue gongzuo 论我国的科学工作 (On our country's scientific work). Beijing: Renmin chubanshe, 1956.

Luo, Pinghan 罗平汉. "1958 nian quanmin daliangang 一九五八年全民大炼钢 (A national campaign of great leap forward in steel in 1958)." *Dangshi wenyuan* 11 (2014): 24–31.

Mao Zedong xuanji 毛泽东选集 (5 vols.). Beijing: Renmin chubanshe, 1967–77.

Mao Zedong zhuzuo xuandu (*shang, xiace*) 毛泽东著作选读（上、下册）. Beijing: Renmin chubanshe, 1986.

Mao, Zuoben 茅左本. *Women zuxian de chuangzao faming* 我们祖先的创造发明 (Inventions of our ancestors). Shanghai: Laodong chubanshe, 1951.

Mao, Zuoben 茅左本. *Women zuxian de chuangzao faming* 我们祖先的创造发明 (Inventions of our ancestors). Shanghai: Renmin chubanshe, 1957.

Marx, Karl, and Friedrich Engels. *The Holy Family or Critique of Critical Criticism Against Bruno Bauer and Company*. Critical Battle against French Materialism (1845), in Marx/Engels Archive. Last accessed December 11, 2018. https://www.marxists.org/archive/marx/works/1845/holy-family/ch06_3_d.htm.

Matsumoto, Toshiro. "Continuity and Discontinuity from the 1930s to the 1950s in Northeast China: The 'Miraculous' Rehabilitation of the Anshan Iron and Steel Company Immediately After the Chinese Civil War." In *The International Order of Asia in the 1930s and 1950s*, edited by Shigeru Akita and Nicholas J. White, 255–73. Surrey: Ashgate, 2010.

Matten, Marc A. "Coping with Invisible Threats: Nuclear Radiation and Science Dissemination in Maoist China." *East Asian Science, Technology and Society* 12, no. 3 (2018): 235–56.

Matten, Marc A. "Turning Away from the Big Brother: China's Search for Alternative Sources of Knowledge During the Sino-Soviet Split." *Comparativ* 29, no. 1 (2019): 64–90.

Medvedjev, Shores A. *The Rise and Fall of T. D. Lysenko*. New York: Columbia University Press, 1971.

Miao, Litian 苗力田. "'Zhishi jiushi liliang'—jinian Fulanxisi Peigen dansheng sibai zhounian '知识就是力量'—纪念弗兰西斯．培根诞生四百周年 (Knowledge is Power—Commemorating the 400th birthday of Francis Bacon)." *Renmin Ribao*, Jan. 22, 1961.

Miller, H. Lyman. *Science and Dissent in Post-Mao China: The Politics of Knowledge*. Seattle: University of Washington Press, 1996.

Myrdal, Jan. *Bericht aus einem chinesischen Dorf*. München: dtv, 1969.

Nedostup, Rebecca. *Superstitious Regimes: Religion and the Politics of Chinese Modernity*. Cambridge: Harvard University Press, 2009.

Needham, Joseph, Tsuen-hsuin Tsien, Dieter Kuhn, Francesca Bray, Christian Daniels, Nicholas K. Menzies, Christoph Harbsmeier, et al., eds. *Science and Civilisation in China* (7 vols.). Cambridge: Cambridge University Press, 1954–2008.

Needham, Joseph. "Science and Society in East and West." *Science and Society* 28, no. 4 (1964): 385–408.

Needham, Joseph. "Science Reborn in China: Rise And Fall of the Anti-Intellectual Gang." *Nature* 274 (1978): 832–34.
Needham, Joseph. "The Roles of Europe and China in the Evolution of Oecumenical Science." *Journal of Asian History* 1, no. 1 (1967): 3–32.
Neibu cankao 内部参考 (Inner References). 1953–1958.
Nongcun qingnianshe 农村青年社. *Shuanglun shuanghuali jianghua* 双轮双铧犁讲话 (Speech on the two-wheel two-blade plow). Beijing: Zhongguo qingnian chubanshe, 1956.
Nongye Jishu 农业技术 (*Agricultural Technology*). 1957–68.
Nongye jixiebu bangongting 农业机械部办公厅, ed. *Quanguo nongye jixie shumu 1949–1960* 全国农业机械书目 1949–1960 (Catalogue of Books on Agricultural Machinery from the Whole Country, 1949–1960). No publisher indicated, 1961.
Nongyebu jixieju 农业部机械局. *Huifu shuanglun shuanghuali de mingyu* 恢复双轮双铧犁的名誉 (Restoring the reputation of the two-wheel, two-blade plow). Beijing: Nongye chubanshe, 1958.
Nongye jixie xuebao 农业机械学报 (*Transactions of the Chinese Society of Agricultural Machinery*). 1957–2010.
Nongye Jixie 农业机械 (*Agricultural Machines*). 1966–71.
Nongye kexue tongxun 农业科学通讯 (*Agricultural Science News*). 1949–59.
Nongye Zhishi 农业知识 (*Agricultural Knowledge*). 1951–83.
Nongyebu xumu shouyiju 农业部畜牧兽医局. *Zhongshouyi yanfang huibian (1956 nian quanguo minjian shouyi zuotanhui ziliao)* 中兽医验方汇编 (1956 年全国民间兽医座谈会资料) (A Collection of Tested Recipes of Chinse Veterinary Medicine: Materials from the 1956 National Forum of Folk Veterinary Medicine). Beijing: Nongye chubanshe, 1957.
Orleans, Leo A. "Soviet Influence on China's Higher Education." In *China's Education and the Industrialized World: Studies in Cultural Transfer*, edited by Ruth Hayhoe and Marianne Bastid, 184–98. Armonk: M. E. Sharpe, 1987.
Orleans, Leo A., ed. *Science in Contemporary China*. Stanford: Stanford University Press, 1980.
Palmer, David A. *Qigong Fever: Body, Science, and Utopia in China*. New York: Columbia University Press, 2007.
Peng, Guanghua 彭光华. "Zhongguo kexuehua yundong xiehui de chuangjian, huodong jiqi lishi diwei 中国科学化运动协会的创建、活动及其历史地位 (Foundation, Activities and Historical Assessment of the Society for the China Scientization Movement)." *Zhongguo keji shiliao* 中国科技史料 13, no. 1 (1992): 60–72.
Perry, Elizabeth J. "The Promise of PRC History." *Journal of Modern Chinese History* 10, no. 1 (2016): 113–17.
Pollock, Ethan. *Stalin and the Soviet Science Wars*. Princeton, NJ: Princeton University Press, 2006.
Pomeranz, Kenneth. *The Great dDivergence: China, Europe and the Making of the Modern World Economy*. Princeton, NJ: Princeton University Press, 2001.
Porter, Theodore M. *Trust in Numbers: The Pursuit of Objectivity in Science and Public Life*. Princeton, NJ: Princeton University Press, 1995.

Qian, Xuesen 钱学森. *Lun renti kexue* 论人体科学 (On somatic science). Beijing: Renmin junyi chubanshe, 1988.

Qian, Ying. "The Shopfloor as Stage: Production Competition, Democracy, and the Unfulfilled Promise of Red Flag Song." *China Perspectives*, no. 2 (2015): 7–14.

Qinbing fangzhi bianxiezu 禽病防治编写组. *Qinbing fangzhi* 禽病防治 (Prevention and Treatment of Poultry Diseases). Shanghai: Renmin chubanshe, 1975.

Qiushi 求是 (*The Seeking*). 1988–today.

Qiu, Hongzhong 邱鸿钟. *Yixue yu renlei wenhua: yixue wenhua shehuixue yinlun* 医学与人类文化: 医学文化社会学引论 (Medicine and Human Culture: A Sociological Introduction to Medical Culture). Changsha: Hunan kexue jishu chubanshe, 1993.

Quanguo kexue dahui wenjian 全国科学会议文件 (*National Science Conference Documents*). Beijing: Renmin chubanshe, 1978.

Quanguo nongju zhanlanhui 全国农具展览会, ed. *Nongju tuxuan* 农具图选 (Selected Blueprints of Farming Tools, vols. 1–20). Beijing: Nongye chubanshe, 1958.

Quanguo nongju zhanlanhui 全国农具展览会, ed. *Quanguo nongju zhanlanhui—tuijian zhanpin—nongtian paiguan jixie* 全国农具展览会—推荐展品—农田排灌机械 (National Exhibition of Agricultural Tools—Recommended Exhibits—Farmland Irrigation Machines). Beijing: Kexue puji chubanshe, 1958.

Quanguo zhongshouyi jingyan xuanbian 全国中兽医经验选编 (Selected experiences of Chinese veterinary medicine). Beijing: Kexue chubanshe, 1977.

Qunzhong kexue yanjiu wenji 群众科学研究文集 (Collected Writings of Scientific Research of the Masses A Collection of [the Result] of Mass Scientific Research). Beijing: Kexue puji chubanshe, 1958.

Renmin Huabao 人民画报 (*People's Pictorial*).

Renmin Ribao 人民日报 (*People's Daily*).

Revolutionary Health Comittee of Hunan Province, ed. *A Barefoot Doctor's Manual*. London: Routledge, 1977.

Rhoads, Robert A., Wang Xiaoyang, Shi Xiaoguang, and Chang Yongcai, eds. *China's Rising Research Universities: A New Era of Global Ambition*. Baltimore: Johns Hopkins University Press, 2014.

Rogaski, Ruth. *Hygienic Modernity: Meanings of Health and Disease in Treaty-Port China*. Berkeley: University of California Press, 2004.

Ruoshui 若水. "Makesi zhuyi de renshilun shi shijian lun 马克思主义的认识论是实践论 (The Epistemology of Marxism Is Practice)." *Renmin Ribao*, Feb. 16, 1963.

Rupnow, Dirk, Veronika Lipphardt, Jens Thiel, and Christina Wessely, eds. *Pseudowissenschaft*. Frankfurt: Suhrkamp, 2008.

Saich, Tony. *China's Science Policy in the 80s*. Atlantic Highlands: Humanities Press International, 1989.

Sanyuan xian kexue jishu weiyuanhui 三原县科学技术委员会, and Sanyuan xian kexue jishu xiehui 三原县科学技术协会. *Zhuwen de zazhen zhiliao* 猪瘟的扎针治疗 (Acupuncture treatment of swine diseases). Xi'an: Shaanxi renmin chubanshe, 1960.

Schäfer, Dagmar, ed. *Cultures of Knowledge: Technology in Chinese History*. Leiden: Brill, 2012.

Scheid, Volker. *Chinese Medicine in Contemporary China: Plurality and Synthesis.* Durham: Duke University Press, 2002.

Scheid, Volker. "Convergent Lines of Descent: Symptoms, Patterns, Constellations, and the Emergent Interface of Systems Biology and Chinese Medicine." *East Asian Science, Technology and Society* 8, no. 1 (2014): 107–39.

Schmalzer, Sigrid. *Red Revolution, Green Revolution: Scientific Farming in Socialist China.* Chicago: University of Chicago Press, 2016.

Schmalzer, Sigrid. *The People's Peking Man: Popular Science and Human Identity in Twentieth-Century China.* Chicago: University of Chicago Press, 2008.

Schmalzer, Sigrid. "Youth and the 'Great Revolutionary Movement' of Scientific Experiment in 1960s–1970s Rural China." In *Maoism at the Grassroots: Everyday Life in China's Era of High Socialism*, ed. Jeremy Brown and Matthew D. Johnson, 154–78. Cambridge: Harvard University Press, 2015.

Schmid, Sonja. "Celebrating Tomorrow Today: The Peaceful Atom on Display in the Soviet Union." *Social Studies of Science* 36, no. 3 (2006): 331–65.

Schneider, Laurence. *Biology and Revolution in Twentieth-Century China.* Lanham: Rowman and Littlefield, 2003.

Schroeder-Gudehus, Brigitte, and David Cloutier. "Popularizing Science and Technology During the Cold War: Brussels 1958." In *Fair Representation: World's Fairs and the Modern World*, edited by Robert W. Rydell and Nancy Gwinn, 157–80. Amsterdam: VU University Press, 1994.

Science for the People. *China: Science Walks on Two Legs—A Report from Science for the People.* New York: Avon Books, 1974.

Selden, Mark. "Mao Zedong and the Political Economy of Chinese Development." *China Report* 24, no. 2 (1988): 125–39.

Selected Works of Mao Tse-tung (5 vols.). Beijing: Foreign Languages Press, 1977.

Shaanxi sheng nongye gongju gaige zhanlanguan 陕西省农业工具改革展览馆. *Xianjin gongju tupu* 先进工具图谱 (Atlas of Advanced Tools). Xi'an: Shaanxi renmin chubanshe, 1960.

Shanghai shi nongyeju 上海市农业局. *Shanghai shi nongye zhanlanhui* 上海市农业展览会 (Agricultural Exhibition in Shanghai). Shanghai, 1957.

Shapiro, Judith. *Mao's War against Nature—Politics and the Environment in Revolutionary China.* Cambridge: Cambridge University Press, 2001.

Shen, Jie 沈洁. "'Fan mixin' huayu jiqi xiandai qiyuan '反迷信' 话语及其现代起源 (The discourse of 'anti-superstition' and its modern origin)." *Shilin* 史林, no. 2 (2006): 34–42.

Shen, Zhihua 沈志华, ed. *Eluosi jiemi dang'an xuanbian—Zhong-Su guanxi* 俄罗斯解密档案选编 中苏关系 (Selected declassified Russian archives: Sino-Soviet relations, 12 vols.). Shanghai: Dongfang chuban zhongxin, 2015.

Shen, Zhihua 沈志华. *Sulian zhuanjia zai Zhongguo* 苏联专家在中国 (Soviet experts in China). Beijing: Xinhua chubanshe, 2009.

Shi, Bainian 史柏年. "1958 nian dalian gangtie yundong shuping 1958 年大炼钢铁运动述评 (A critical review of the 1958 Great Leap Forward in Iron and Steel)." *Zhongguo jingjishi yanjiu* 中国经济史研究 2 (1990): 124–33.

Shi, Zhe 施哲. *Shijianlun he ren de zhengque sixiang shi cong nali laide qianshuo* 《实践论》和《人的正确思想是从那里来的》浅说 (A brief discussion of "On Practice" and "Where do correct ideas come from"). Changchun: Jilin renmin chubanshe, 1975.

Shinn, Terry, and Richard Whitley, eds. *Expository Science: Forms and Functions of Popularisation*. Dordrecht: Lancaster, 1985.

Shiqipansiji 施契潘斯基 (W. Shepanskii). *Gongchang zhong qunzhong zhengzhi gongzuo* 工厂中群众政治工作 (The masses' political work in factories), translated by Lin Xiu 林秀. Shanghai: Shidai chubanshe, 1950.

Sidel, Victor W. "The Barefoot Doctors of the People's Republic of China." *New England Journal of Medicine*, no. 24 (1972): 1292–1300.

Siegelbaum, Lewis H. "1929: Shock Workers." Seventeen Moments in Soviet History. Accessed November 27, 2020. http://soviethistory.msu.edu/1929-2/shock-workers.

Siegelbaum, Lewis H. *Stakhanovism and the Politics of Productivity in the USSR, 1935–1941*. New York: Cambridge University Press, 1988.

Siemens, Johannes. "Lyssenkoismus in Deutschland (1945–1965)." *Biologie in unserer Zeit*, no. 27 (1997): 255–62.

Sivin, Nathan. *Chinese Alchemy: Preliminary Studies*. Cambridge: Harvard University Press, 1968.

Sivin, Nathan. *Traditional Medicine in Contemporary China: A Partial Translation of Revised Outline of Chinese Medicine (1972) with an Introductory Study on Change in Present-day and Early Medicine*. Ann Arbor: Center for Chinese Studies, University of Michigan, 1987.

Sivin, Nathan. "Why the Scientific Revolution Did Not Take Place in China—or Didn't It?" Last accessed January 18, 2021. http://ccat.sas.upenn.edu/~nsivin/writ.html (revised version of an essay published 1982 in *Chinese Science*, no. 5 [2005]: 45–66).

Sleeboom-Faulkner, Margaret. *The Chinese Academy of Science (CASS): Shaping the Reforms, Academia and China*. Leiden: Brill, 2007.

Smith, Arthur H. *Chinese Characteristics*. New York: Fleming H. Revell, 1894.

Smith, Steven. "Introduction: The Religion of Fools? Superstition: Past and Present." *Past and Present* 199 (2008): 7–55.

Smith, Steven A. "Local Cadres Confront the Supernatural: The Politics of Holy Water (*Shenshui*) in the PRC, 1949–1966." *China Quarterly*, no. 188 (2006): 999–1022.

Smith, Steven A. "Talking Toads and Chinless Ghosts: The Politics of 'Superstitious' Rumors in the People's Republic of China, 1961–1965." *The American Historical Review* 111, no. 2 (2006): 405–27.

Stakhanov, Alekseĭ Grigor'evich (А. Г. Стахáнов 斯達哈諾夫). *Sidahanuofu yundong* 斯达哈诺夫运动 (The Stakhanovite Movement). Translated by Sun Siming 孫斯鳴. Shanghai: Shanghai shidai shuju, 1949.

Stavis, Benedict. *The Politics of Agricultural Mechanization in China*. Ithaca: Cornell University Press, 1978.

Strauss, Julia C. "Rethinking Land Reform and Regime Consolidation in the People's Republic of China: The Case of Jiangnan (1950–1952)." In *Rethinking China in the 1950s*, edited by Mechthild Leutner, 24–34. Berlin: Lit Verlag, 2007.

Strong, Anna Louise. *The Rise of the People's Communes in China*. New York: Marzani & Munsell, 1960.

Sulian jiachu fanyu de xianjin lilun he jingyan 苏联家畜繁育的先进理论和经验 (Advanced Theory and Experiences of the Soviet Union's Livestock Breeding). Beijing: Caizheng jingji chubanshe, 1956.

Sulian Nongye Kexue 苏联农业科学 (*Soviet Agricultural Sciences*). 1951–58.

Tao, Xingzhi 陶行知. *Shenghuo jiaoyu wenxuan* 生活教育文选 (Selected Works on Life Education), edited by Hu Xiaofeng 胡晓风. Chengdu: Sichuan jiaoyu, 1988.

Taylor, Kim. *Chinese Medicine in Early Communist China, 1945–1963: A Medicine of Revolution*. London: Routledge, 2005.

Ter Haar, Barend J. "China's Inner Demons: The Political Impact of the Demonological Paradigm." *China Information* XI, nos. 2/3 (1996–97): 54–88.

"The Conflict Between Mao Tse-tung and Liu Shao-chi over Agricultural Mechanization in Communist China." *Current Scene* 6, no. 17 (1968): 9.

Ullerich, Curtis. *Rural Employment & Manpower Problems in China*. New York: M. E. Sharpe, 1979.

Urbansky, Sören, and Arunabh Ghosh. "Introduction." *The PRC History Review* 2, no. 3 (June 2017): 1–3.

Wang, Chonglun 王崇伦, and Xu Binzhang 许彬章. *Rang women he shijian saipao* 让我们和时间赛跑 (Let's compete with time). Beijing: Gongren chubanshe, 1954.

Wang, Jiankun 王建坤, and Zhang Qinghua 张清华. *Gailiang nongju de qiaomen* 改良农具的窍门 (*The Know-how of Improving Farming Tools*). Beijing: Zhongguo qingnian chubanshe, 1958.

Wang, Lunxin 王伦信. "Minguo shiqi de gonggong kexueguan yu zhongxue like shiyan jiaoxue 民国时期的公共科学馆与中学理科试验教学 (Public science halls in Republican China and the teaching of experiment in middle school science education." *Journal of South China Normal University,* no. 10 (2007): 89–94.

Wang, Yuhu 王毓瑚. *Zhongguo xumushi ziliao* 中国畜牧史资料 (Collection of Materials on the History of Animal Husbandry in China). Beijing: Kexue chubanshe, 1958.

Wang, Zikun 王梓坤. "Shitan ziran kexue yanjiu de yiban fangfa 试谈自然科学研究的一般方法 (On the general methods of natural science research)." *Hongqi*, no. 2 (1979): 55–62.

Wang Zuoyue. "Saving China through Science: The Science Society of China, Scientific Nationalism, and Civil Society in Republican China." *Osiris*, 2nd Series, 17, Science and Civil Society (2002): 291–322.

Wang, Zuoyue. "The Chinese Developmental State During the Cold War: The Making of the 1956 Twelve-year Science and Technology Plan." *History and Technology* 31, no. 3 (2015): 180–205.

Watson, James. "Feeding the Revolution: Public Mess Halls and Coercive Commensality in Maoist China." In *Handbook of Food and Anthropology*, edited by James Watson and Jakob Klein, 308–20. London: Bloomsbury Press, 2016.

Wei, Chunjuan Nancy, and Darryl E. Brock, eds. *Mr. Science and Chairman Mao's Cultural Revolution: Science and Technology in Modern China*. Lanham: Lexington Books, 2013.

Wilson, James, and James Keeley. *China, the Next Science Superpower? The Atlas of Ideas: Mapping the New Geography of Science*. London: Demos, 2007.

Wu, Youxun 吴有训. *Zhongguo Kexueyuan wulixue shuxue huaxue bu baogao* (1955 nian 6 yue 2 ri zai Zhongguo Kexueyuan xuebu chengli dahuishang de baogao) 中国科学院物理学数学化学部报告 [1955年6月2日在中国科学院学部成立大会上的报告] (Report on the Department of Physics, Mathematics and Chemistry of the Chinese Academy of Sciences [Report at the founding meeting of the Chinese Academy of Sciences on June 2, 1955]). *Lun woguo de kexue gongzuo* 论我国的科学工作 (Discussing our country's scientific work. Beijing: Renmin chubanshe, 1956.

Xie, Guoxian 谢国贤 and Tao Lüxiang 陶履祥. *Jiachu binglixue zong lun* 家畜病理学总论 (General Discussion on Livestock Pathology). Shanghai: Shangwu yin shuguan, 1951.

Xie, Yutong 谢裕通. *Jiachu yibing shouce* 家畜疫病手册 (Handbook on livestock plagues). Shanghai: Zhonghua shuju, 1951.

Xiong, Yuezhi 熊月之. *Xixue dongjian yu wan Qing shehui* 西学东渐与晚清社会 (Western Learning Spreading to the East and Late Qing Society). Shanghai: Renmin chubanshe, 1994.

Xuexi Angang jishu gexin jingyan 学习鞍钢技术革新经验 (Studying AnSteel's experiences in technological innovation). Chongqing: Chongqing renmin chubanshe, 1954.

Xu, Guohua 许国华, and Liu Jiansheng 刘健生. *Tuolaji he lianhe shougeji* 拖拉机和联合收割机 (Tractor and combine harvester). Beijing: Zhonghua quanguo kexue jishu puji xiehui, 1954.

Xu, Liangying 许良英, and Fan Dainian 范岱年. *Kexue he woguo shehuizhuyi jianshe* 科学和我国社会主义建设 (Science and Socialist Construction in Our Country). Beijing: Renmin chubanshe, 1957.

Xu, Liangying, and Fan Dainian. *Science and Socialist Construction in China*. Armonk: M. E. Sharpe, 1982.

Xu, Xiaoqun. "'National Essence' vs 'Science': Chinese Native Physicians' Fight for Legitimacy, 1912–37." *Modern Asian Studies* 31, no. 4 (1997): 847–77.

Xu, Xiaxiang 徐霞翔. "Toushi nongcun dianying fangyingyuan—yi ershi shiji wushi niandai Jiangsu sheng wei li 透视农村电影放映员—以二十世纪五十年代江苏省为例 (The film project team in rural areas—taking Jiangsu Province of the 1950s as an example)." *The Twentieth Century* (Web version), published March 31, 2009, last access December 20, 2020, http://www.cuhk.edu.hk/ics/21c/media/online/0812018.pdf.

Xue kexue 学科学 (*Studying Science*). 1956–66.

Xun, Lu (Lu Hsün). *Complete Stories* (transl. by Yang Xianyi). Bloomington: Indiana University Press, 1981.

Yang, Dali. "State and Technological Innovation in China: A Historical Overview, 1949–89." *Asian Perspective* 14, no. 1 (1990): 91–112.

Yang, Jisheng. *Tombstone: The Untold Story of Mao's Great Famine*. London: Allen Lane, 2012.

Yang, Li 杨力, Gao Guangyuan 高广元, and Zhu Jianzhong 朱建中. *Zhongguo kejiao dianying fazhan shi* 中国科教电影发展史 (History of Development of Science Education Films in China). Shanghai: Fudan University Press, 2010.

Yang, Nianqun, "The Memory of Barefoot Doctor System." In *Governance of Life in Chinese Moral Experience: The Quest of an Adequate Life*, edited by Everett Zhang, Arthur Kleinman, and Tu Weiming, 131–45. London: Routledge, 2011.

Yang, Nianqun 杨念群. *Zaizao "bingren": Zhong-Xi yi chongtuxia de kongjian zhengzhi* 再造"病人": 中西医冲突下的空间政治 (Remaking "Patients": Spacial Politics in the Conflicts between Chinese and Western Medicine). Beijing: Zhongguo Renmin daxue chubanshe, 2013.

Yang, Ruisong 楊瑞松. *Bingfu, huanghuo yu shuishi: "Xifang" shiye de Zhongguo xingxiang yu jindai Zhongguo guozu lunshu xiangxiang* 病夫、黃禍與睡獅: "西方" 視野的中國形象與近代中國國族論述想像 (Sick man, yellow peril and sleeping lion: the image of China in the eyes of the West and the narrative imagination of the modern Chinese nation). Taibei: Zhengda chubanshe, 2010.

Yang, Wenzhong 杨文仲. "Xin zhongguo de zhonggongye 新中国的重工业 (Heavy industry of the new China)." In *Xin zhongguo huifu shiqi de zhonggongye jianshe* 新中国恢复时期的重工业建设 (Heavy industry construction in the recovery period of the new China), edited by Yang Wenzhong, 1–15. Beijing: Sanlian shudian, 1954.

Yejin gongye chubanshe 冶金工业出版社. *Zai Angang xianfa de guanghui qizhi xia qianjin* 在鞍钢宪法的光辉旗帜下前进 (Marching under the glorious flag of the AnSteel Charter). Beijing: Yejin gongye chubanshe, 1975.

Yejinbao 冶金报 (*Journal of Metallurgy*). 1956.

Yichuan yu yuzhong 遗传与育种 (*Heredity and Breeding*). 1975–78.

Yixue yu zhexue 医学与哲学 (*Medicine and Philosophy*).

Yong tu banfa shixian ban jixiehua 用土办法实现半机械化 (Using local methods to achieve half-mechanization). Nanjing: Jiangsu renmin chubanshe, 1958.

Yu, Benyuan 喻本元, and Yu Benheng 喻本亨. *Chongbian jiaozheng Yuan Heng liaoma niu tuo jing quanji* 重编校正元亨疗马牛驼经全集. Beijing: Nongye chubanshe, 1963.

Yu, Benyuan 喻本元, and Yu Benheng 喻本亨. *Yuan Heng liaoma ji: Fu niu tuo jing* 元亨疗马集: 附牛驼经. Beijing: Nongye chubanshe, 1957.

Yu, Guangyuan 于光远. *Zhongguo de kexue jishu zhexue: ziran bianzhengfa* 中国的科学技术哲学: 自然辩证法 (China's science, technology and philosophy: natural dialectics). Beijing: Kexue chubanshe, 2013.

Yu, Priscilla C. "Leaning to One Side: The Impact of the Cold War on Chinese Library Collections." *Libraries & Culture* 36, no. 1 (2001): 253–66.

Yunnan sheng nongyeju 云南省农业局, and Xumu shouyi kexue yanjiusuo 畜牧兽医科学研究所. *Zhongshouyi jingyan xuanji (shang)* 中兽医经验选辑 (上) (Selected Experiences of Chinese Veterinary Medicine). No publisher indicated, 1973.

Zedong, Mao 毛泽东. *Jingji wenti yu caizheng wenti* 经济问题与财政问题 (On Economic and Financial Questions). Dongbei shuju, 1948.

Zedong, Mao 毛泽东. "Ren de zhengque sixiang shi cong nali laide 人的正确思想是从哪里来的？ (Where do correct ideas come from? 1963)." In *Mao Zedong zhuzuo xuandu (xiace)* 毛泽东著作选读（下册）, 839–41. Beijing: Renmin chubanshe, 1986.

Zhang, Dai 张岱. *Zenyang zai gong kuang qiye shixing yizhangzhi* 怎样在工矿企业中实行一长制 (How to implement the one-man-management system in industry and mining?). Beijing: Gongren chubanshe, 1954.

Zhang, Qiyun 张其昀. "Kexue yu kexuehua 科学与科学化 (Science and scientization)." *Kexue de Zhongguo* 1, no. 1 (1933): 4.

Zhang, Qiyun 张其昀. "Zhongguo kexuehua yundong xiehui faqi zhiqu shu 中国科学化运动协会发起旨趣书 (The Objectives of the Association for the China Scientization Movement)." *Kexue de Zhongguo* 科学的中国 1, no. 1 (1933): 1–3.

Zhang, Shengwen 张声文. "Chijiao shouyi wei renmin 赤脚兽医为人民 (Barefoot veterinary doctor for the people)." *Guangxi nongye kexue* 广西农业科学, no. 5 (1976): 40–41.

Zhang, Tianlai 张天来, et al. *Zai kexue jinjun de daolu shang* 在科学进军的道路上 (On the road of marching towards science). Shenyang: Liaoning renmin chubanshe, 1956.

Zhishi jiushi liliang 知识就是力量 (*Knowledge Is Power*). 1956–62.

Zhonggong Shanghai shiwei nongcun gongzuo weiyuanhui nongju gaige bangongshi 中共上海市委农村工作委员会农具改革办公室, ed. *Shanghai shi xinshi nongju xuanji* 上海市新式农具选辑 (*Selection of new farming tools in Shanghai City*). Shanghai: Kexue jishu chubanshe, 1960.

Zhonggong zhongyang wenxian yanjiushi 中共中央文献研究室. *Jianguo yilai zhongyao wenxian xuanbian* 建国以来重要文献选编 (20 vols.). Beijing: Zhongyang wenxian chubanshe, 1992–2011.

Zhonggongye tongxun 重工业通讯 (*Heavy Industry Newsletter*). 1953–56.

"Zhongguo gongchandang dashiji. 1956 nian 中国共产党大事记. 1956年." *News of the Communist of China* 中国共产党新闻. Accessed November 15, 2018. http://cpc.people.com.cn/GB/64162/64164/4416035.html.

Zhongguo jingji lunwenxuan bianji weiyuanhui 中国经济论文选编辑委员会. *Xiang Sulian zhuanjia xuexi* 向苏联专家学习 (*Learning from the Soviet Experts*). Beijing: Sanlian shudian, 1953.

Zhongguo kexueyuan dongwu yanjiusuo 中国科学院动物研究所. *Zhubing fangzhi shouce* 猪病防治手册 (Pamphlet of preventing and treating swine diseases). Beijing: Kexue chubanshe, 1979.

Zhongguo Nongken 中国农垦 (*Statefarms in China*). 1957–64.

Zhongguo nongye baike quanshu 中国农业百科全书 (Chinese agriculture encyclopedia, 25 vols.). Beijing: Nongye chubanshe, 1986–96.

Zhongguo nongye kexueyuan zhongshouyi yanjiusuo 中国农业科学院中兽医研究所. *Zhongshouyi zhenjiuxue* 中兽医针灸学 (Acupuncture and moxibustion in Chinese veterinary medicine). Beijing: Nongye chubanshe, 1959.

Zhongguo nongye kexueyuan zhongshouyi yanjiusuo 中国农业科学院中兽医研究所. *Zhongshouyi zhenduanxue* 中兽医诊断学 (Diagnostics in Chinese Veterinary Medicine). Beijing: Nongye chubanshe, 1972/1962.

Zhongguo renmin jiefangjun shouyi daxue 中国人民解放军兽医大学. *Zhongshouyi yanfang ji* 中兽医验方集 (Collected Recipes in Chinese Veterinary Medicine). Beijing: Jiefangjun shouyi daxue, 1972.

Zhongguo renmin zhengzhi xieshang huiyi gongtong gangling 中国人民政治协商会议共同纲领 (Common Program of Chinese People's Political Consultative Conference)." *Jiangxi zhengbao* 江西政报 (Jiangxi Government Information), no. 3 (1949): 19–20.

Zhongguo xumu shouyi xuehui ed. 中国畜牧兽医学会. *Zuguo youliang jiachu pinzhong* 祖国优良家畜品种 (The motherland's fine livestock breeds). Beijing: Kexue chubanshe, 1956.

Zhonghua quanguo zonggonghui shengchanbu 中华全国总工会生产部. *Shengchan jingsai wenti jianghua* 生产竞赛问题讲话 (Speeches on the Problems of Production Competition). Beijing: Gongren chubanshe, 1950.

Zhongguo xumu shouyi xuehui zhu chuanranbing zhuanye yanjiuzu 中国畜牧兽医学会猪传染病专业研究组, *Zhu de zhuyao chuanranbing jiqi fangzhi fangfa* 猪的主要传染病及其防治方法 (Main infectious diseases of pigs and their prevention methods) (Beijing: Caizheng chubanshe, 1956).

Zhongshouyi yiyao zazhi 中兽医医药杂志 (*Journal of Traditional Chinese Veterinary Medicine*).

Zhongyang jiaoyu kexue yanjiusuo 中央教育科學研究所. *Lao jiefangqu jiaoyu ziliao—kangri zhanzheng shiqi* 老解放區教育資料－抗日戰爭時期 (Materials on education of the liberated bases—the war of resistance against Japan period). Beijing: Jiaoyu kexue, 1986.

Zhongyang renmin kexueguan 中央人民科学馆. *Kang Mei yuan Chao yundong zhong de dongbei yu Chaoxian tuji* 抗美援朝运动中的东北与朝鲜图集 (Atlas of the Northeast and Korea in the Campaign of Resisting America and Aiding Korea). Shanghai: Zhonghua shuju, 1951.

Zhou, Enlai 周恩来. "Guowuyuan guanyu jiaqiang minjian shouyi gongzuo de zhishi 国务院关于加强民间兽医工作的指示 (1956)." In *Ningxia shouyizhi* 宁夏兽医志 (The Gazetteer on Veterinary Medicine in Ningxia), edited by Zhou Shengjun 周生俊, 341–43. Ningxia: Ningxia renminchubanshe, 2012.

Zhou, Shangwen 周尚文, Li Peng 李鹏, and Hao Yuqing 郝宇青. *Xin Zhongguo chuqi "liu-Su chao" shilu yu sikao* 新中国初期"留苏潮"实录与思考 (Documents and thoughts about the "waves of studying in the Soviet Union" in early People's Republic of China). Shanghai: Huadong shifan daxue chubanshe, 2012.

Zhu, Shine 祝世讷. *Zhongyi wenhua de fuxing* 中医文化的复兴 (Rejuvenation of Chinese medicine culture). Nanjing: Nanjing chubanshe, 2013.

Zhu, Xianling 朱显灵 and Hu Huakai 胡化凯. "Shuanglun shuanghuali yu zhongguo xinshi nongju tuiguang gongzuo 双轮双铧犁与中国新式农具推广工作 (Work to promote the two-wheel two-blade plow and Chinese new farming tools)." *Dangdai Zhongguoshi yanjiu* 16, no. 3 (2009): 56–63.

Zhubing fangzhi bianxiezu, ed. 猪病防治编写组. *Zhubing fangzhi* 猪病防治 (Prevention and treatment of swine diseases). Shanghai: Shanghai renmin chubanshe, 1972.

Index

Academy of Chinese Medical Sciences, xiv, 105
acupuncture, 106, 109–10, 113–16, 126, 132; and moxibustion, 109, 113–16; in animal husbandry, 114; points, 114
All–China Association for the Dissemination of Scientific and Technological Knowledge, 34, 36, 48, 62
animal husbandry, 59–60, 62, 106, 109, 111–14. *See also* acupuncture
AnSteel (Anshan Iron and Steel Works), 42, 82, 84, 93–96; Charter, 83, 95–96
Anti–Rightist Campaign, 16, 66
artifact, 38–40, 42
Association for Veterinary Medicine, 105. *See also* veterinary medicine
Association of China's Education Film, 43

Bacon, Francis, xx
ball–bearing, 61, 67–70, 126
barefoot doctor, 106, 114; for veterinary medicine, 106, 114; veterinary, 115. *See also* veterinary medicine
Beijing Agriculture University, 107–109
biomedicine, 4, 6, 10, 105–106, 112, 115–16, 133

bourgeois, 3, 19, 33, 83, 129; class, 13, 19; expert, 127
bureaucratism, xx

Campaign against Spiritual Pollution, 130
Campaign of Technological Innovation 84
CCP (Chinese Communist Party), xv, xvii–xviii, xix–xx, xxiii, 6–7, 12, 14–15, 19, 29, 31–32, 34, 36, 59–60, 67, 82, 93, 105
Chen, Duxiu, 4
Chen, Lifu, 43
Chen, Wangdao, 35
Chen, Zhengren, 63
China Association for Science and Technology, 34
China Scientization Movement, 1, 30–32, 43; Association for, 1, 30. *See also* Chen, Lifu; Zhang, Qiyun
Chinese medicine. *See* medicine (medical)
Chinese veterinary medicine. *See* medicine (medical)
class, xvii, xxii, 11, 13, 15, 18–19, 21, 33, 36, 59, 81, 85, 87, 113–14, 117; politics, xv, xix, 35, 51; social, xiv, 17, 81, 128; struggle, xx–xxi, xxiii,

17–19, 34, 51, 59, 111, 128–30; working, 47, 81, 84, 86, 90, 93, 126, 128–29
cognitive gap, 39, 47
combine harvester, 62, 64
contingency, 87, 91; experiment without, xix–xxiii; historical, 134
cooperative, 64–65, 67
cost, 29, 35, 43, 64–65, 69, 84, 87–89, 92, 106, 114–16
Cultural Revolution, xv, xvii–xviii, xxii–xxiii, 84, 113–14, 126–29, 131–32

data, xvii–xviii, xxi, 11, 89, 91. *See also* statistics
Deng, Xiaoping, xxiii, 126, 128, 130, 132
Dewey, John 21
diagnosis, 3, 4, 50, 109, 110, 113, 116
dianxing. *See* typical case
Dong, Chuncai, 32

Eight-Point Charter of Agriculture, 60
epistemic, 60, 111; multiplicity, 127; rupture and continuity, 126–31; space, 117
epistemics, 132
epistemological, 11, 19, 71, 96, 105–106, 110–11, 114, 127; categories, 70; chaos, 39; flexibility, 134; traditions, 134; transformation, 2
Eurocentrism, 3
exhibition, 31–33, 37, 38–43, 44–45, 49, 69, 70–71, 86, 92–94, 111
experience (*jingyan*), xiii–xv, xvii–xxii, 5, 11–13, 21, 29, 33–39, 44, 48, 51, 70, 83, 94–95, 107–108, 110–12, 116–17, 127, 132–34; empirical, xv, xix, xxii, 51, 72; local, 12, 114; practical, 17
experiment (*shiyan* 试验, *shiyan* 实验): scientific, xxi, xxiii, 91, 127, 130. *See also* experimental plot; practice
experimental plot, xx, 70

fact, xviii, 3, 11, 19, 37–38, 40
Fang, Yi, 130
First National Conference on Public Health, 14
Five-Year Plan, 14, 60, 62–63, 65, 83
folk veterinary practitioner. *See* veterinary medicine
formula. *See* prescription
Forty Articles of Agriculture, 65
Four Modernizations, xxiii, 127

Gangtie. *See Steel and Iron*
Gao, Shiqi, 32, 35
Gu, Chaohao 36
Gu, Junzheng, 35

He, Zuoxiu, 133
health, 4–9, 34, 44, 47, 106–107, 133
hog cholera, 107, 115
Hu, Qiaomu, 128
Hu, Sheng, 130
Hu, Shi, 2–4, 21, 134
Hua, Luogeng, 37
hygiene, 6, 9, 38, 116, 133

idealism, xxii, 3, 113
internal reading material, 62, 74
innovation, 15, 17, 33, 37, 47, 60–61, 68–69, 81, 85, 94, 125–27; campaign of technological, 84; exhibition of technological, 94; technical and technological, 44–45, 47, 83–87, 91, 92, 94, 96

Jiefang Ribao. *See Liberation Daily*
Jixie gongren. *See Machine Workers*
Journal of Metallurgy, 87, 91

Kexue dazhong. *See Popular Science*
Kexue huabao. *See Science Pictorial*
Kexue puji gongzuo. *See Science Dissemination Work*
Kexue puji tongxun. *See Newsletter of Science Dissemination*
Khrushchev, Nikita, 15

knowledge:
  communication, 9, 127; plurality of, 15, 134; production, xix, xxii, 12, 17–18, 29–30, 35–37, 51, 60–61, 81–83, 86–87, 93, 95–96, 106, 117, 126–28, 131–32, 134–35; systems, xv, xix, 33, 134; veterinary, 105–108, 110–11, 116. *See also* science (scientific)
Knowledge is Power, xx, 47–51, 88, 90
Korean War, 35, 37, 40, 42, 84

labor, xvii, 14, 45, 48, 62–64, 69, 82, 85, 90, 116, 125, 129–30, 132
laboring people, xiii–xv, xix, 33–35, 37
Latour, Bruno, 10
Liberation Daily, 12, 31
Lin, Biao, xviii, 96, 113–114, 126
Liu, Shaoqi, 64, 66, 72, 96, 113–114
Liu, Xianzhou, 18
Lu, Dingyi, 15, 21, 129
Lysenko, Trofim, 16, 25, 130

Ma, Chengde, 94
Machine Workers, 85
management, 14, 59, 64, 72, 81–83, 85, 93, 95–96
Marx, Karl, xx
Marxism, 117, 129–130, 133
Marxism–Leninism, xiii–xiv, xix–xx, 5, 13–14, 16, 32, 34, 50–51
mass science, xviii, 16–18, 29, 105, 126
mass–line, 17, 83, 86, 96, 112, 114, 127, 129, 131
MaSteel Charter, 96
materialism, xx, xxii, 5, 13–14, 33, 38–39, 51, 62, 117
mechanization, xvii, 5, 17, 51, 59, 61–72, 81, 92, 113, 125
medicine (medical):
  Chinese, xiii–xiv, 3–6, 33, 105–108, 116, 132–33, 139; National Medicine Movement, 5; Chinese veterinary (CVM), 51, 105, 110, 112–117; worker, xiv, 9, 105; Western, xv, 5–6, 105–106, 112, 114, 116, 125, 132
Minjian shouyi tongxun. *See Newsletter of Folk Veterinary Medicine*
Ministry of Agricultural Machinery, 63
mixin. *See* superstition, superstitious practice
model laborer/worker, 14, 81, 84, 93
modernity. *See* socialist modernity
modernization, xix, xxiii, 4–7, 12, 15–16, 30, 32, 35, 45, 50, 60, 62, 81, 96–97, 125, 129, 134; accelerated, 14, 38, 82; project of, 94, 116–117. *See also* Four Modernizations
"more, faster, better, and more economical," 17, 87
moxibustion, 109. *See also* acupuncture

National Conference on Research in Chinese and Western Veterinary Medicine, 110
National Science Conference, 126, 128
native, xiii, 12, 16, 29. *See also tu* (indigenous)
native metals:
  iron, 82, 88–92; nodular cast iron, 88–90; steel, 92
Natural Science Popularization Movement, 13, 31
Newsletter of Folk Veterinary Medicine, 108
Newsletter of Heavy Industry, 85, 87, 94
Newsletter of Natural Dialectics, 129–30
Newsletter of Science Dissemination, 32, 34, 37
Nie, Rongzhen, 17, 110

objective facts/objectivity, xvi, xix, 11, 32, 37, 40, 71, 131
On Contradiction, 97
On Practice, xx–xxii, 97, 132
On the Problem of Intellectuals, 14

On the Ten Major Relationships, 15, 62
One–man management (edinonachalie), 82–83

Patriotic Production Competition, 84
patriotism, 1, 132
Peng, Zhen, 67
People's Daily, xx–xxii, 6, 14, 17, 37–38, 46, 62, 65–67, 86, 88, 92, 94, 96, 107–108, 114, 130
People's Pictorial, 87
Popular Science, 47, 68, 75
practice. *See* experience (*jingyan*); experiment (*shiyan* 试验, *shiyan* 实验); *On Practice*
prescription, 22, 109–112
prevention, 6, 112–113, 116
production:
  agricultural, xx, 6, 44, 59–61, 64, 107, 114, 125; competition, 69, 82–84, 94; economic, xx–xxii, 13, 15, 17–19, 33–35, 37, 44, 50, 81–83, 89–90, 93, 96, 110, 116; industrial, 81–83, 87, 126
productivity:
  increase, xvii, 16–18, 33–34, 59–60, 62–65, 84–85, 96, 105
progress, 32, 48, 86, 126–28, 130, 134
Prospective Plan for the Development of Science and Technology between 1956 and 1967, 15
pseudoscience, 19, 132–33

Qian, Sanqiang, 130
Qian, Xuesen, 132

rationality, xvi, 2, 4, 11, 71, 131
Red Flag, 17, 19, 130–31
Renmin Huabao. *See People's Pictorial*
Renmin Ribao. *See People's Daily*
resources:
  constraints of, xix, 5, 34, 60, 67, 94; human, xix, 6, 29, 83, 93; local, 69, 132; natural/material, 6, 17, 29, 42–43, 83

science (scientific):
  as primary productive forces, 126, 128; colonial, 16; definition of, xv, 1, 12–13, 19, 71; discourse of, xv, xix–xx, 11, 19, 30, 37, 126–27, 131; dissemination, 1, 70, 131–32; history of, xiv, xvi–xvii, xx, 10–12, 127, 133–134; *Marching towards*, 15, 48, 95; mass, xviii, 16–18, 29, 105, 126; natural sciences, 13–14, 30, 113; People's Science 14, 19, 29, 33, 35, 37, 40, 45–47, 51; popular, xiv, xv, 29, 31–33, 35–36, 47–48, 87–88, 93; production and circulation of, xvi; proletarian, 129; standards, xv–xvi; taxonomies of, xiv, xvi, 38, 112. *See also* exhibition; knowledge; pseudoscience
Science Dissemination Work, 34, 35
Science Pictorial, xiii, 47, 50, 91, 92
Scientific China, 1
Shanghanlun, 10
Sick Man of East Asia, 3
Smith, Arthur, 2
socialist:
  agriculture, 63; construction, 7, 9, 61; culture, 36; distribution of renumeration, 82, 84; emulation, 64, 83–84; future, 48; genetics, 16; industrialization, 15, 82, 93; modernity, xxii, 15, 81; political system, 29, 35, 37, 66, 116, 126; science, 16–18, 48
Soviet Agricultural Science, 62
Stakhanovite Movement, 83–84, 86; Stakhanov, Alekseĭ G., 83
Stalin, Joseph, 16, 61, 83–84, 87, 96
Stalinism, 81–82
statistics, xviii, xxi, 40, 42, 91. *See also* data
Steel and Iron, 87
Sun, Yat–sen, 4

superstition, superstitious practice, xiv, xix, 2–7, 18–20, 69, 72, 105–106, 108, 111, 116, 129, 132

Tan, Jiazhen, 16
Tan, Zhenlin, 67
Tao, Xingzhi, 31
technician, 37, 43, 45–46, 64, 68, 85, 94–96, 114, 127, 129
technocratic experts, technocrats, 64, 72, 87–88, 90, 129, 131–32
technology:
  agricultural, 59, 61, 72; indigenous, 61; industrial, xix, 95; intermediate, 67
tractor. *See* mechanization
tool improvement, 61, 72, 84, 93
tu (indigenous) xvii, 17, 43, 60–61, 69, 71, 88. *See also* native
typical case, xix, xxi, 45, 60, 64, 67–68, 71

veterinary medicine:
  barefoot doctor for, 106, 114; Chinese (CVM), 51, 105, 110, 112, 116. *See also* Association for Veterinary Medicine; medicine (medical)
voluntarism, xxiii, 64, 91, 126

"walking on two legs," 17, 89
Wang, Chonglun, 81, 84–86
Wang, Daofu, 108
Wang, Daohan, 88–89
Wang, Yuhu, 109–10
Wang, Zikun, 129–30
"white steel," 82, 92–93, 132

Xiong, Dashi (T. S. Hsiung), 108–109

yang (foreign), xvii, 18, 60–61
Yejinbao. See Journal of Metallurgy
Yu, Benheng, 109
Yu, Benyuan, 109
Yu, Guangsheng, 88–89
Yu, Guangyuan, 128–29
Yu, Yunxiu, 4, 6, 133
Yue, Tianyu, 107

Zedong, Mao, xiii–xv, xx–xxii, 3, 12, 15, 18–19, 62–63, 65, 68, 83, 87, 95, 96, 105–106, 125, 132
Zeng, Xisheng, 17
Zhang, Qiyun, 1, 30
Zhang, Zhongjing, 10
Zhishi jiushi liliang. See Knowledge is Power
Zhonggongye tongxun. See Newsletter of Heavy Industry
Zhongshouyi. See Chinese veterinary medicine
Zhou, Enlai, 14, 66, 72, 86, 107, 129
Zhou, Jianren, 35
Zhou, Yang, 14, 21

About the Authors

Marc Andre Matten, Ph.D. (2007), is Professor of Contemporary Chinese History at Friedrich-Alexander University Erlangen-Nuremberg, Germany. He has published on the issues of Chinese nationalism and national identity, including *Places of Memory in Modern China—History, Politics, and Identity* (Brill, 2012) and *Imagining a Postnational World—Hegemony and Space in Modern China* (Brill, 2016), as well as on questions on the transnational history of science in Maoist China, such as in the special issue "Moving Knowledge—The Soviet Union and China in the Twentieth Century" of the journal *Comparativ* (1/2019, ed. together with Julia Obertreis).

Rui Kunze is currently a DFG-funded research fellow at the University of Erlangen-Nuremberg, Germany. Her work on the history of science and science fiction in modern and contemporary China has appeared in *East Asian History*, *Twenty-First Century* (二十一世紀), and the edited volume *Chinese Visions of Progress, 1895–1949* (Brill).

www.ingramcontent.com/pod-product-compliance
Lightning Source LLC
Chambersburg PA
CBHW061715300426
44115CB00014B/2704